MATERIALS SCIENCE AND ENGINEERING

FORGING STRONGER LINKS TO USERS

Committee on Materials Science and Engineering:
Forging Stronger Links to Users

National Materials Advisory Board
Commission on Engineering and Technical Systems

National Research Council

Publication NMAB-492

NATIONAL ACADEMY PRESS
Washington, D.C.

NATIONAL ACADEMY PRESS • 2101 Constitution Avenue, NW • Washington, DC 20418

NOTICE: The project that is the subject of this report was approved by the Governing Board of the National Research Council, whose members are drawn from the councils of the National Academy of Sciences, the National Academy of Engineering, and the Institute of Medicine. The members of the panel responsible for the report were chosen for their special competencies and with regard for appropriate balance.

This study by the National Materials Advisory Board was conducted under a contract with the Department of Defense, National Aeronautics and Space Administration, National Science Foundation, and the Department of Energy. Any opinions, findings, conclusions, or recommendations expressed in this publication are those of the author(s) and do not necessarily reflect the view of the organizations or agencies that provided support for the project.

Available in limited supply from:
National Materials Advisory Board
2101 Constitution Avenue, N.W.
Washington, D.C. 20418
202-334-3505
nmab@nas.edu

Additional copies are available for sale from:
National Academy Press
Box 285
2101 Constitution Ave., N.W.
Washington, D.C. 20055
800-624-6242
202-334-3313 (in the Washington
  metropolitan area)
http://www.nap.edu

International Standard Book Number: 0-309-06826-6

Library of Congress Catalog Card Number 99-68968

Cover: Scanning electron micrograph of a device with IBM's six-level copper interconnect technology. Source: Courtesy of International Business Machines Corporation. Unauthorized use not permitted.

THE NATIONAL ACADEMIES

National Academy of Sciences
National Academy of Engineering
Institute of Medicine
National Research Council

The **National Academy of Sciences** is a private, nonprofit, self-perpetuating society of distinguished scholars engaged in scientific and engineering research, dedicated to the furtherance of science and technology and to their use for the general welfare. Upon the authority of the charter granted to it by the Congress in 1863, the Academy has a mandate that requires it to advise the federal government on scientific and technical matters. Dr. Bruce M. Alberts is president of the National Academy of Sciences.

The **National Academy of Engineering** was established in 1964, under the charter of the National Academy of Sciences, as a parallel organization of outstanding engineers. It is autonomous in its administration and in the selection of its members, sharing with the National Academy of Sciences the responsibility for advising the federal government. The National Academy of Engineering also sponsors engineering programs aimed at meeting national needs, encourages education and research, and recognizes the superior achievements of engineers. Dr. William A. Wulf is president of the National Academy of Engineering.

The **Institute of Medicine** was established in 1970 by the National Academy of Sciences to secure the services of eminent members of appropriate professions in the examination of policy matters pertaining to the health of the public. The Institute acts under the responsibility given to the National Academy of Sciences by its congressional charter to be an adviser to the federal government and, upon its own initiative, to identify issues of medical care, research, and education. Dr. Kenneth I. Shine is president of the Institute of Medicine.

The **National Research Council** was organized by the National Academy of Sciences in 1916 to associate the broad community of science and technology with the Academy's purposes of furthering knowledge and advising the federal government. Functioning in accordance with general policies determined by the Academy, the Council has become the principal operating agency of both the National Academy of Sciences and the National Academy of Engineering in providing services to the government, the public, and the scientific and engineering communities. The Council is administered jointly by both Academies and the Institute of Medicine. Dr. Bruce M. Alberts and Dr. William A. Wulf are chairman and vice chairman, respectively, of the National Research Council.

Dedication

T HIS REPORT IS DEDICATED TO THE MEMORY OF ROBERT LAUDISE, the chairman of the NMAB at the time this report was commissioned and a prime mover in developing the theme of the report. We all remember Robert as a person of uncommon technical ability, with the vision and passion required of a leader; yet so warm and human, with a happy smile and quick wit. We shall miss him greatly.

> Blessed is he who carries within himself a God, an ideal and who obeys it—ideal in art, ideal in science, ideal in gospel virtues; therein lies the springs of great thoughts and great actions: they all reflect light from the infinite.
>
> Louis Pasteur (1822–1895)

Preface

MATERIALS ARE IMPORTANT. NEW MATERIALS often provide opportunities for rapid technological advancements, but to seize those opportunities, the materials must be adapted and integrated into economically viable products. As history shows, this has not been easy. Studies show that it often takes 20 or more years for a new material to make a significant penetration into the market. Many challenges will have to be overcome for the nation to derive the full benefit of new materials essential for a vibrant, safe, and environmentally friendly economy. The materials community has an opportunity to play a central role, but it will require changes in both mind-sets and methods.

Can the leisurely pace be improved? Probably. This report examines the many links in the chain from basic research to the introduction of a new material into the market and discusses how the links can be strengthened to accelerate the introduction of new materials into the marketplace. Many factors influence the effectiveness of these interactions, including maturity of the industry, frequency of major changes in the product, openness to innovation, profitability, and competitiveness. As a consequence, new materials find their way from the laboratory to the marketplace by a multitude of pathways.

Although no single formula can ensure the rapid introduction of new materials to the marketplace, practices and policies that facilitate the introduction of new materials have been identified. The objective of this report is to broaden the understanding of the complex factors that can impede the introduction of new materials and to suggest changes in practices and policies to promote the introduction of new materials: researchers must have a better understanding of the constraints of the marketplace; users must be more receptive to new materials and processes; and educators must focus more attention on team building, industrial ecology, design, and production. Most important, a way must be found to navigate the so-called "valley of death," the transfer of the materials technology from the researcher to the end-user.

For this report, the committee conducted in-depth studies of three industry sectors: the automotive industry, the jet-engine industry, and the computer-chip and information-storage industries. In addition to the expertise of the committee members, the committee conducted workshops for each case-study industry. Representatives of the MS&E community, the industrial research community, supply companies, and systems integrators participated in the workshops. The information gathered in these workshops was synthesized and used as a basis for this report and the development of findings and recommendations.

Comments and suggestions can be sent via electronic mail to nmab@nas.edu or by FAX to NMAB (202) 334-3718.

Dale F. Stein, chair
Committee on Materials Science and Engineering:
Forging Stronger Links to Users

Acknowledgments

T HE COMMITTEE WOULD LIKE TO THANK THE PRESENTERS and participants in the three industry workshops that served as the principal data-gathering sessions for this study. Presenters at the November 1997 electronics industry workshop were: Barry Schechtman, NSIC; Sheldon Schultz, University of California at San Diego; David Thompson, IBM Almaden Research Center; Robert Rottmayer, Read-Rite; Thomas Howell, Quantum; Paul Peercy, SEMI/ SEMATECH; Woodward Yang, Harvard University; Don Shaw, Texas Instruments; Alain Harrus, Novellus; Pier Chu, Motorola; James McElroy, NEMI; Michael Pecht, University of Maryland; William Chen, IBM; Jack Fischer, Interconnection Technology Research Institute; and Robert MacDonald, Intel. Presenters at the January 1998 turbine-engine industry workshop were: Ambrose Hauser, GE Aircraft Engines; Michael Goulette, Rolls-Royce PLC; Gary Roberge, Pratt and Whitney; Anthony Giamei, United Technologies; Kathy Stevens, Wright Laboratories; James Williams, GE Aircraft Engines; Peter Shilke, GE; Harry Brill-Edwards, consultant; Robert Noel, Ladisch Company, Inc.; Greg Olson, QuestTek Innovations; Gernant Maurer, Special Metals Corporation; Ken Harris, Cannon-Muskegon Corporation; William Parks, U.S. Department of Energy; Mark Fulmer, Federal Aviation Administration; Tony Freck, consultant; and Larry Fernbacher, Technology Assessment and Transfer. Presenters at the March 1998 automotive industry workshop were: Christopher Magee, Ford Motor Company; Roger Heimbuch, General Motors; Sam Harpest, Honda; Andrew Sherman, Ford Motor Company; Darryl Martin, AISI International; Peter Bridenbaugh, Alcoa; Ken Rusch, Budd Plastics; Kenneth Browell, GE; John McCracken, TWN, Inc.; John Allison, Ford; Floyd Buch, Duralcan; Joseph Heremans, General Motors; Sam Froes, University of Idaho; Bryan McEntire, Norton; and Jeff

Dieffenbach, IBIS Associates. The committee would like to give special thanks to Ivan Amato for developing the case study vignettes.

This report has been reviewed by individuals chosen for their diverse perspectives and technical expertise, in accordance with procedures approved by the NRC's Report Review Committee. The purpose of this independent review is to provide candid and critical comments that will assist the authors and the NRC in making the published report as sound as possible and to ensure that the report meets the institutional standards for objectivity, evidence, and responsiveness to the study charge. The content of the review comments and draft manuscript remain confidential to protect the integrity of the deliberative process. We wish to thank the following individuals for their participation in the review of this report: Kathleen Taylor, General Motors; Robert Aplan, Pennsylvania State University; Gernant Maurer, Special Metals Corporation; Robert Eagan, Sandia National Laboratories; Robert Pfahl, Motorola; Julia Weertman, Northwestern University; and James Williams, Ohio State University. While the individuals listed above have provided many constructive comments and suggestions, responsibility for the content of this report rests solely with the authors.

Finally, the committee gratefully acknowledges the support of the staff of the National Materials Advisory Board, including Robert Schafrik, former director of the NMAB, who got the project off the ground; Robert Ehrenreich, who provided technical support and program management throughout the data-gathering and report-development phases of the study; Thomas Munns, who shepherded the report through review and publication; and Pat Williams, who provided administrative support throughout the entire study.

Contents

Tables, Figures, and Boxes

xv

BOXES

Acronyms

ATP	Advanced Technology Program
CMOS	complementary metal oxide semiconductors
CRADA	cooperative research and development agreement
DARPA	Defense Advanced Research Projects Agency
DOD	U.S. Department of Defense
DOE	U.S. Department of Energy
ERC	engineering research centers
FAA	Federal Aviation Administration
GMR	giant magnetoresistance
GOALI	Grant Opportunities for Academic Liaison with Industry
HDD	hard disk drive
I/O	input/output
IOF	Industries of the Future
I/UCRC	industry/university cooperative research centers
ManTech	Manufacturing Technology Program
MR	magnetoresistance
MS&E	materials science and engineering
NEMI	National Electronics Manufacturing Initiative
NAICS	North American Industrial Classification System
NSIC	National Storage Industry Consortium
NSTC	National Science and Technology Council
OEM	original equipment manufacturer
PNGV	Partnership for a New Generation of Vehicles
R&D	research and development
SIC	Standard Industrial Classification
S/IUCRC	state/industry university cooperative research centers
TWG	technical working group

Executive Summary

THIS IS A REPORT ABOUT RELATIONSHIPS—how to understand them and how to nurture them. What relationships? The relationships among the producers of materials and the users of materials. These relationships are depicted visually in Figure ES-1.

This is also a report about processes. What processes? The processes of innovation—from the generation of knowledge through development and application to the ultimate integration of a material into a useful product. These processes are typically and linearly depicted in Figure ES-2.

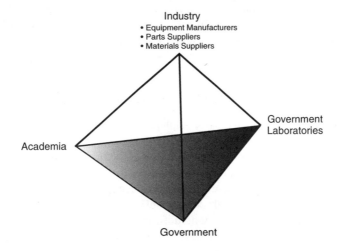

FIGURE ES-1 Relationships in the MS&E community.

1

Phase 0: Knowledge-Base Research	Phase 1: Material Concept Development	Phase 2: Material/Process Development	Phase 3: Transition to Production	Phase 4: Product Development

FIGURE ES-2 Notional phases of the innovation process.

The committee recognized, however, that no two developments are alike, and development processes are actually a series of iterative decision loops. The real-life case studies in this report provide a more accurate depiction of the actual nonlinear processes between the generation of new knowledge and the integration of materials into useful products (Chapter 2).

The purpose of the report is to recommend ways to strengthen the linkages among the key participants in the materials science and engineering (MS&E) community to accelerate the rate at which new ideas are integrated into finished products. The entire process can now take decades. Continuing U.S. competitiveness requires that the time be shortened. The report describes the relationships and incentives of those who affect the MS&E community. The report then recommends how those relationships could be strengthened to accelerate the rate of introduction of new materials into the economy.

The committee drew on experiences with three distinctive MS&E applications—advanced aircraft turbines, automobiles, and computer chips and information-storage devices. The committee examined these industries *not* to provide in-depth descriptions and evaluations of the linkages in these particular industries but to gain the insights from these industries to support general propositions about strengthening linkages in the MS&E community as a whole.

The committee's recommendations reinforce many observations, concerns, and recommendations being made in many different forums about the creative processes of research and development (R&D) and the importance of supporting them in a climate of budget reductions in both the public and private sectors (NRC, 1999a). The focus of the report, however, is on the relatively young discipline of MS&E.

FINDINGS AND RECOMMENDATIONS

The committee concluded that a complete definition of MS&E must incorporate materials categories (e.g., metals, polymers, ceramics, composites), functionally differentiated end-use categories (e.g., electronic materials, biological materials, structural materials), functional interrelationships (i.e., structure, properties, processing, and performance), as well as needs and constraints of users of materials throughout the materials value chain. To capture these complexities, the committee developed the following definition of MS&E.

To extend the usefulness of all classes of materials, the field of MS&E seeks to understand, control, and improve upon five basic elements:

- the life-cycle **performance** of a material in an application (i.e., in a component or system)
- the design and **manufacture** of a component or system, taking advantage of a material's characteristics
- the **properties** of a material that make it suitable for manufacture and application
- the **structure** of a material, particularly as it affects its properties and utility
- the **synthesis** and **processing** by which a material is produced and its structure established

The committee believes that the MS&E community should serve the near-term and long-term needs of the ultimate users of products made from materials. Therefore, the fundamental focus of this report is the importance of materials advances in the development of marketable products. The successful commercialization of materials and process advances is generally driven by one of four end-user forces: (1) cost reduction; (2) cost-effective improvement in quality or performance; (3) societal concerns, manifested either through government regulation or self-imposed changes to avoid government regulation; or (4) crises.

Substantially different forces drive the MS&E R&D communities: (1) the availability of funding; (2) expansion of the basic knowledge base; (3) fulfillment of an educational mission; (4) the desire for professional recognition; and (5) the availability of equipment. A new material/process is not likely to be researched by the MS&E R&D community and adopted by industry unless it satisfies at least one of the perceived needs of each community.

Detailed recommendations to improve linkages between the MS&E and the end-user communities throughout the materials/process development and commercialization processes are included in Chapter 3. Although all of these recommendations are important, the committee found that overcoming the barriers to Phase 2 (material/process development) R&D is the most promising way to shorten the time to market of laboratory innovations. Phase 2, or the "valley of death," is the transition point at which materials/process innovations change from a "technology push" from the MS&E research community to a "product pull" from the end-user community. The committee recommends that the MS&E and user communities focus their efforts on strengthening linkages during this phase of technology development.

Despite major differences between industries, some general approaches can be taken to improve Phase 2 R&D. The key to accelerating the passage through Phase 2 is to establish an environment in which (1) innovations are desired and anticipated by those who will use them and (2) business considerations are addressed early in the development process by researchers. The committee believes that focusing on improving the Phase 1 linkages that set the stage for product pull and establishing the potential viability of an industry for Phase 3 and Phase 4 (getting down to business) will improve the chances that materials and processing innovations will be successfully commercialized.

The committee recommends that the following primary mechanisms be given priority to establish product pull in the early stages of technology development (during Phase 1 and, perhaps, Phase 0):

- consortia and funding mechanisms to support "precompetitive" research
- industry road maps to set priorities for materials research
- university centers of excellence to coordinate multidisciplinary research and facilitate industry-university interactions

The committee recommends that the following developments be given priority to improve the transition of materials advances from Phase 2 to production implementation:

- collaboration with end-user industries to identify the type of data required by product designers to assess new material/processes
- investigation of methods to improve the research infrastructure for materials suppliers and parts suppliers
- extension of the patent-protection period, especially for applications that require extended certification periods
- development of industrial ecology as an integral part of the education and expertise of both MS&E researchers and product designers
- development of a regulatory climate based on constructive cooperation and goal setting to promote the adoption of new materials that achieve or enhance societal goals

REPORT ORGANIZATION

The findings of the committee have been organized into four chapters. Chapter 1 describes the importance of materials and processing technology to the U.S. economy, develops a taxonomy to bound the field of MS&E, and describes the study task. Chapter 2 introduces the conceptual schema the committee used to assess the materials development and commercialization processes. Chapter 3 contains the committee's analysis of the critical linkages between industry, government, and universities and recommends ways to improve these linkages to accelerate the commercialization of new materials and processes. Finally, Chapter 4 contains priority recommendations for improving the materials development process and reducing the time to market of advances in materials and process technologies.

REFERENCE

NRC (National Research Council) 1999a. Harnessing Science and Technology for America's Economic Future. Washington, D.C.: National Academy Press.

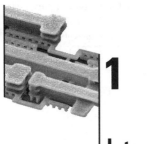

1

Introduction

THE OVERALL INTENT OF THE FIELD of materials science and engineering (MS&E) is to enable the production of components and systems to improve the function, effectiveness, efficiency, and economy of products and thereby enhance the quality of life and standard of living for humankind. MS&E cuts across both the science and engineering of materials and the relationships of matter and its use. MS&E deeply affects all segments of life, from society at large to industry to the global environment.

The National Science and Technology Council (NSTC) stated in 1995 that advanced materials were the foundation and fabric of manufactured products. To support their assertion, NSTC cited the role of advanced materials in providing robust structures for fuel-efficient automobiles and damage-resistant buildings, enabling electronic devices that can transmit signals rapidly over long distances, protecting bridges and other surfaces from wear and corrosion, and endowing jet engines and airframes with sufficient strength and heat tolerance to permit supersonic flight. The NSTC concluded that many leading commercial products and military systems could not exist without advanced materials and that many of the new products critical to the nation's continued prosperity will come to be only through the development and commercialization of advanced materials (NSTC, 1995).

Although it is difficult to quantify, materials make a significant contribution to the economy. According to data compiled by the U.S. Department of Commerce, the value of industry shipments of basic raw materials (Standard

TABLE 1-1 Value of Industry Shipments of Basic Raw Materials in 1996

SIC Code	Description	Value of Shipments (billions)
2821	Plastic materials and resins	$40.1
2824	Organic fibers	$12.9
331	Blast furnace and basic steel products	$74.5
333	Primary nonferrous metals	$15.4

Source: DOC, 1998.

Industrial Classification[1] [SIC] codes 2821, 2824, 331, and 333) amounted to approximately $143 billion in 1996 (Table 1-1). This value significantly underestimates the contribution of materials to the economy, however. Because of the vital role of MS&E in the processing of materials, another approach to estimating the contribution of materials to the domestic economy would be to include all parts formed by a single fabrication technology, such as casting, molding, forging, or stamping. Adding the value of the shipments for materials-intensive manufactured products in SIC codes 282, 30, 32, 33, and 34, the contribution of materials to the U.S. economy is roughly $685 billion (Table 1-2). This figure overestimates the contribution of materials because manufacturing costs are included in the total. An average of these estimates of upper and lower bounds yields a value of about $400 billion.

It could be argued that even $685 billion understates the contribution of materials to the economy because a modern economy (and much of the $3.8 trillion manufacturing sector [DOC, 1999]) could not exist without materials. Advances in MS&E have enabled improvements in many sectors of the economy. For example, the materials components of complex manufactured systems (e.g., jet engines, automobiles, and computer-chip and information-storage computer components) are not included in these data.

It is generally agreed that the United States leads the world in materials research and development (R&D), especially the development of advanced materials (NAS, 1998). Nevertheless, many are concerned that the United States does not lead the world in the commercialization of advanced materials. The objective of the committee convened by the National Materials Advisory Board of the National Research Council that conducted this study was to determine changes

---

[1] Standard Industrial Classifications were replaced with North American Industrial Classification System (NAICS) for the 1997 Economic Census. NAICS codes for basic raw materials include 3311 (Iron and Steel Mills and Ferroalloy Manufacturing), 3313 (Alumina and Aluminum Production and Processing), 3314 (Nonferrous Metal [except Aluminum] Production and Processing), and 3252 (Resin, Synthetic Rubber, and Artificial and Synthetic Fibers and Filaments Manufacturing). Data for 4-digit NAICS are incomplete at press time.

TABLE 1-2 Value of Industry Shipments of Fabricated Raw Materials

Industry Classification			Value of Shipments (billions)	
SIC	NAICS	Description	1996	1997
282	3252	Plastic materials and synthetics	$59.6	NA
30	326	Plastics and rubber products manufacturing	$150.5	$159.0
32	327	Nonmetallic mineral product manufacturing	$82.4	$88.3
33	331	Primary metal manufacturing	$178.3	$192.9
34	332	Fabricated metal product manufacturing	$214.0	$233.7

Source: DOC, 1998, 1999.

both within the MS&E and end-user communities that would facilitate the adoption of new materials, reduce the number of "missed opportunities," and improve interactions between materials end-users and the MS&E community. This report focuses on the linkages between materials R&D and the commercialization of materials and suggests ways to promote the introduction of advanced materials into the marketplace to ensure that the United States maintains its leadership in industrial sectors that depend on materials.

TAXONOMY

One of the most daunting aspects of any study of the MS&E discipline is defining the field. Although materials and processes have fueled technological progress for thousands of years, the field of MS&E *per se* did not exist prior the 1960s. The designation of MS&E as a single discipline arose from the coalescence of three previously distinct, materials-specific fields. The roots of MS&E as a discipline are grounded most directly in the fields of metallurgy, ceramics, and polymer science. Although many other disciplines (e.g., physics, geology, electronics, optics, chemistry, and biology) overlap with MS&E and have made indispensable contributions to its development as a formal discipline, these three materials-based disciplines are at the heart of the origin of MS&E. A practical definition of the field is the study of science and engineering principles related to the discovery and understanding, production, use, recycling, and disposal of materials.

An alternative definition was put forth in *Material Science and Engineering for the 1990s: Maintaining Competitiveness in the Age of Materials* (NRC, 1989). Rather than defining the field by classifying materials by categories, this definition focused on the common elements of the MS&E discipline, regardless of material type (Figure 1-1). Formatted as a tetrahedron, MS&E is defined as the interrelationships among structure/composition, properties, performance, and

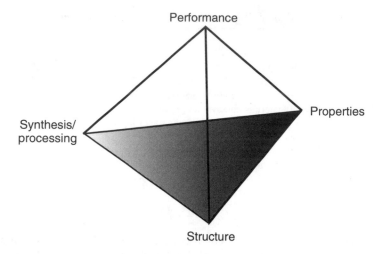

FIGURE 1-1   Graphical representation of the connections among the common elements in the MS&E R&D discipline (independent of material type). Source: NRC, 1989.

synthesis/processing for all types or forms of matter. Advanced carbon fibers, for example, arguably have an extraordinary *property*: their very high Young's modulus, which is a measure of intrinsic stiffness. The modulus of carbon fibers is directly and inextricably linked with the method used to produce them (i.e., their *processing*). The processing defines the fiber's *microstructure*, from which its extraordinary stiffness is derived. Thus, the interrelationships between the property, structure/composition, and processing ultimately determine the *performance* of carbon fiber when it is incorporated into the skin of a fighter jet, the shaft of a golf club, or the spar of a sailboat.

Although the utility of materials developments was considered by the study committee, one limitation of this 1990-vintage tetrahedron is that it does not convey the importance of utility. Thus, it shows no explicit link to the users of materials or the ultimate beneficiaries of MS&E research. The concerns of the MS&E discipline appear to be limited to the four corners of the tetrahedron and doing something useful with materials becomes someone else's responsibility (e.g., a designer, a marketing person, or an entrepreneur). Thus, the MS&E mission of the early 1990s was focused on the pursuit of fundamental scientific and engineering information rather than on finding or assisting in the development of valuable new uses for this information. As a result, information tended to be gathered with little regard for its eventual utility, and users needs did not significantly influence the direction of R&D. Too often, new material systems appeared to have little or no foreseeable user value or potential for production scale-up.

The absence of links joining the MS&E R&D community and materials users should raise concerns with both groups because materials are more than a scientific curiosity. In fact, they are fundamentally important to commerce and society. However, because raw materials *per se* are a commodity for which there is rarely a direct end-use demand, their extrinsic value is difficult to assess. Most consumers do not generally use steel or polyethylene for their own sake. The demand for materials is derived from the demand for the goods in which they are used.

The term "commercialization" implies one of two possibilities: either the embodiment of a technology must be sold in a way that is both profitable and sustainable, without corporate or government subsidies, or it must be used in a component or system that is similarly sold. In short, to be considered a "commercial" product, normal transactions in the market involving its manufacture and sale must result in someone making a profit (NRC, 1997). Commercial considerations are critical because the links between MS&E and ultimate end-users must pass through as many as a half dozen intermediaries, all of whom have needs, requirements, and constraints that must be satisfied. For example, end-user industries have been reducing their product development cycle times in order to increase their competitiveness. Thus, materials developers at the raw-material production stage might also have to reduce their development cycle times to meet the needs of end-user industries.

Based on the broad interests of the MS&E community, which extend all the way from the extraction, synthesis, and refining of a material to its end use and disposal/recycling, a definition of the MS&E community must explicitly link the community with its users. A complete description of MS&E must incorporate materials categories (e.g., metals, polymers, ceramics, composites), functionally differentiated end-use categories (e.g., electronic materials, biological materials, structural materials), functional interrelationships (i.e., structure, properties, processing, and performance), as well as the user needs and constraints throughout the materials value chain (e.g., extraction, synthesis, refining, parts making, systems integration, end-use, and recycling or disposal). In order to try to capture these complexities, the definition for MS&E established by the NRC in 1989 should be revised as follows.

To extend the usefulness of all classes of materials, the field of MS&E seeks to understand, control, and improve upon five basic elements:

- the life-cycle **performance** of a material in an application (i.e., in a component or system)
- the design and **manufacture** of a component or system, taking advantage of a material's characteristics
- the **properties** of a material that make it suitable for manufacture and application
- the **structure** of a material, particularly as it affects properties and utility
- the **synthesis** and **processing** by which a material is produced and its structure determined

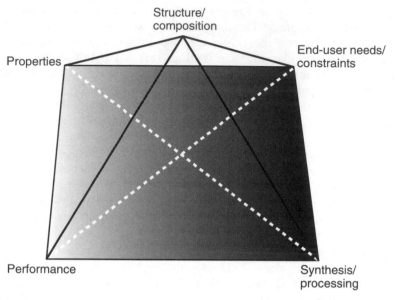

FIGURE 1-2   Graphical representation of the connections among the common elements in the entire MS&E discipline, including the end-user (independent of materials type).

An updated version of the graphic defining MS&E, linking the needs and constraints of the users of materials with the common elements of MS&E is shown in Figure 1-2. MS&E should serve the near-term and long-term needs of the ultimate users of products. These needs should influence the direction of MS&E R&D, whether basic or applied, short term or long term.

STUDY MODE OF OPERATION

To determine changes in both the MS&E and end-user communities that would facilitate the adoption of new materials, reduce the number of "missed opportunities," and improve interactions between materials end-users and the MS&E community, the committee conducted in-depth studies of three industry sectors: the automotive industry, the jet-engine industry, and the computer-chip and information-storage industries. In addition to the expertise of the committee members, the committee invited representatives of the MS&E community, the industrial research communities, the supply companies, and the systems integrators for each of the case-study industries to attend workshops and to share their expertise with the committee. The goals of the workshops were to determine (1) user needs and business practices that promote or restrict the incorporation of materials and processes innovation, (2) the manner in which priorities in materials selection are determined, (3) mechanisms to improve links between the materials community and the engineering disciplines, and (4) programs (e.g., education, procedures, information technology) that could improve these linkages. Summaries of the workshops are provided in Appendices A, B, and C. The information gathered in these workshops was synthesized by the committee and used as a basis for this report and the recommendations.

2

Materials Development and Commercialization Process

A LTHOUGH THE IMPORTANCE OF MATERIALS to the national economy and the profits of individual companies is clear, the timelines and processes by which materials are developed and introduced are harder to characterize. Our poor understanding of these processes is not from lack of study, however. Numerous attempts have been made to define the materials development and commercialization processes (e.g., NRC 1989, 1993, 1997). The problem is that the process rarely follows a linear progression through time from basic research to final implementation. Rather, as the vignettes of this chapter and Chapter 3 show, the development of each and every material can seem to be a singular and unique sequence of activities.

The first step in analyzing the linkages among the MS&E and end-user communities and identifying potential methods for strengthening these linkages to accelerate the implementation of laboratory discoveries is establishing a baseline definition of the materials development and commercialization processes. This chapter presents an overview of (1) the time and drivers for successful transition, (2) a conceptual schema of the transitions from research concept to product integration, and (3) a description of the characteristics of each phase in the conceptual schema. The committee used the simplest and broadest possible view of materials development and commercialization processes for the analysis of linkages, even though it is not applicable to any specific development. Also, because the development of new commercial materials and processes are in most cases inextricably entwined, *material/process* will be referred to as a joint innovation in the remainder of this report.

DURATION AND DRIVERS OF MATERIALS TRANSITIONS

One premise of this study is that materials innovations have traditionally taken a relatively long time to transpire. This premise is based on retrospective substitution analyses for the adoption of some materials. For example, according to a formalized method for tracking these transformations developed by Fisher and Pry (1971), a complete material/process substitution requires at least a decade and, more typically, as much as 25 years (Table 2-1). The cases reviewed during the three workshops organized by the committee supported this conclusion. The following reasons were most often cited for delays in the implementation of new materials/processes:

- *industrial culture*, including aversion to risk, with asymmetrical consequences (e.g., enormous penalties for failure, lesser rewards for success); tradition (e.g., reluctance to change established paradigms); and perceived adequacy of existing technologies
- *industrial infrastructure*, including capital investment in the current technology; narrow, periodic windows of opportunity that can be easily missed by purveyors of new technologies; fragmented structure of industry, which

TABLE 2-1 Examples of Takeover Times and Substitution Midpoints

Substitution	Takeover Time[a] (years)	Substitution Midpoint[b] (year)
Rubber: natural to synthetic	59	1956
Fibers: natural to synthetic	58	1969
Leather: natural to plastic	57	1957
Butter: natural to margarine	56	1957
Specialty steels: open-hearth to electric-arc	47	1947
House paint: oil-based to water-based	43	1967
Steel: Bessemer to open-hearth	42	1907
Turpentine: tree-tapped to sulfate	42	1959
Paint pigment: PbO-ZnO to TiO_2	26	1949
Residence floors: hardwood to plastic	25	1966
Pleasure boat hulls: other to plastic	20	1966
Insecticides: inorganic to organic	19	1966
Tire fibers: natural to synthetic	17.5	1948
Cars: metal to plastics	16	1982
Steels: open-hearth to basic oxygen furnace	10.5	1960
Soap (U.S.): natural to detergent	8.75	1951
Soap (Japan): natural to detergent	8.25	1962

Source: Fisher and Pry, 1971.
[a] time required to progress from 10-percent substitution to 90-percent substitution
[b] when substitution is 50-percent complete

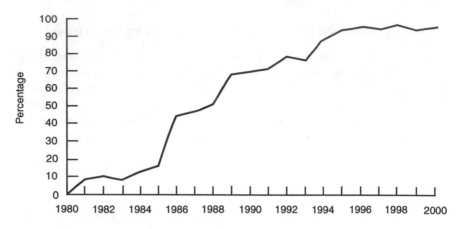

FIGURE 2-1   Timeline for the adoption of single-crystal, first-stage, high-pressure turbine blades for jet engines. Source: Howmet International, Inc.

prevents new technologies from moving up the value chain; incompatibility of new technology with existing system constraints; and ability to improve existing technologies

- *economic issues*, including unreliability of supply (e.g., insufficient availability of materials or insufficient supplier capabilities) and economies of scale (i.e., insufficient volume to justify adoption of a new technology, even if reliable processes and suppliers exist)
- *inadequate support or incomplete knowledge base*, including discontinuity of the development cycle (e.g., termination or suspension of funding); lack of a champion in industry; limited potential of the new technology to meet *all* user requirements; lack of an information or knowledge base for the new technology; and uncertainty of cost modeling (i.e., potential discrepancies between actual cost and theoretical cost determined by modeling)

Successful materials substitutions tend to follow a classic "S"-shaped curve. Figure 2-1, for example, shows the process for the adoption of single-crystal, first-stage turbine blades for jet engines, which enabled engines to operate at higher temperatures and thus more efficiently (Box 2-1). The motivation for changing from the incumbent material-based system to a new system varies for different applications, however. For intake manifolds, for example, plastic versions were less expensive than the previous die-cast aluminum forms. For computer chips, copper interconnects allowed faster processing capabilities (Box 2-2). Based on the numerous examples of successful and unsuccessful material innovations described by the industrial representatives at the workshops, the committee was able to identify four general driving forces that are common to successful implementations.

Finding 2-1. Successful materials developments and transitions will generally be driven by one of four end-user forces: (1) cost reduction (e.g., polymeric trim and intake manifolds in automobiles); (2) improvement in quality or performance or the customers' perceptions of quality or performance (e.g., advanced microprocessors, titanium golf clubs, aluminum wheels in automobiles); (3) societal concerns, manifested either through government regulation or actions to avoid government regulation (e.g., introduction of high-strength steels to reduce automobile weight and thus help meet fuel economy regulations; replacement of chlorofluorocarbons); or (4) crises (e.g., adoption of thermal barrier coatings by South African Airways and synthetic rubber during World War II).

Finding 2-2. Anecdotal evidence at the workshops also revealed significant differences between the forces that drive end-user communities and those that drive academic MS&E R&D communities. Based on comments by academic representatives at the workshops, the committee was able to identify five driving forces that underlie the development of a successful academic R&D program: (1) availability of funding; (2) expansion of the basic knowledge base; (3) fulfillment of an educational mission; (4) desire for professional recognition; and (5) availability of equipment. In general, the MS&E academic community has been unable or unwilling to conduct an R&D program unless at least one of these driving forces is present.

Finding 2-3. The differences between the forces driving the end-users and those driving the MS&E community will determine the eventual success or failure of a materials innovation (Box 2-3). A new material/process is not likely to be researched by the academic MS&E R&D community or adopted by industry unless it satisfies at least one of the perceived needs of both communities.

CONCEPTUAL SCHEMA

Materials development and commercialization processes are extraordinarily complex. Most case studies of materials commercialization are retrospective, starting with successful innovations and tracing their history back to their origins. This hindsight view makes the progressions appear more logical and coherent than they actually are (Holton et al., 1996) and ignores the lengthy process of incremental improvement that continues long after a material/process is initially adopted. In reality, material/process innovation is less a linear progression from basic research to final implementation than a mixture of activities, some of which may be either conducted concurrently or bypassed entirely. For example, Box 2-4 describes the development of tungsten filaments for light bulbs. In this

BOX 2-1
Single-Crystal Turbine Blades

The efficiency of turbine engines, like the ones that propel jet aircraft or generate electrical power, increases about 1 percent for every increment of 12°F in the fuel inlet temperature. This relationship has driven innovations in engine materials that can be used safely at ever-higher temperatures.

The development of turbine blades and vanes made of superalloys that retain their strength even when heated to 90 percent of their melting temperature—unlike conventional metals that weaken at cooler temperatures—has been one the most important factors in keeping engine inlet temperatures hot. But even superalloys have weaknesses. At high temperatures, they become susceptible to grain boundary creep, which makes metal components vulnerable to the massive centrifugal forces generated as the parts whirl around at 25,000 rotations per minute inside the engine. Engine temperatures must be kept low enough so that the blades do not creep toward the engine casing, a scenario that could end in engine damage at least, and a catastrophic accident at worst. The addition of elements like carbon, boron, and zirconium to the superalloy composition strengthens the grain boundaries, but also lowers the alloy's melting temperature, which limits the safe operating temperature.

In the early 1960s, researchers at Pratt and Whitney decided to pursue another approach. Instead of trying to strengthen problematic grain boundaries, they thought it would be even better to eliminate the grain boundaries altogether. The inspiration came, in part, from the burgeoning silicon crystal industry, which was churning out salami-sized single crystals for the nascent microelectronics industry. Their first goal was to eliminate the grains most susceptible to grain boundary creep under the stress conditions caused by the centrifugal forces in operating engines. By the mid-1960s, they had achieved their goal through processing innovations. One key was keeping the bottom of the ceramic mold much cooler than the top and letting the zone of cooling and solidification rise through the molten metal very slowly—over the course of hours. The result was blades with columnar grains instead of grains with boundaries in all directions (equiaxed). This process, which became known as directional solidification, increased high-temperature strength by several hundred percent.

This success suggested that performance could be improved further by eliminating the remaining grain boundaries. The technical key to creating a single-crystal turbine blade was to develop a "crystal selector" that would allow a single grain to grow into the bottom of the ceramic mold and then grow outward and upward to fill the mold as a single crystal. The first single-crystal turbine blades were in hand by the end of the 1960s. Yet single-crystal turbine blades did not appear on the commercial market until 1982. Part of the delay was due to the failure of some early directionally-solidified blades in a military test, which eroded confidence in the new technology. This failure led to an intense, five-year research effort by Pratt and Whitney to address the remaining metallurgical problems.

Even so, directionally solidified and single-crystal blades cost more to produce than the easier-to-make equiaxed blades because the production process has a substantially lower yield of usable material. The production process involves not

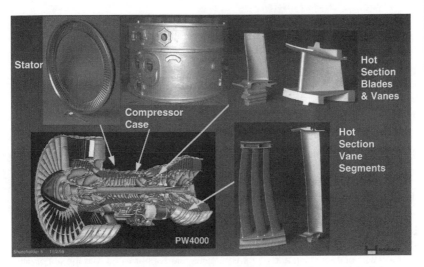

Turbine engine components. Photo provided by Howmet International, Inc.

only complicated metallurgy, but also sophisticated ceramic technology and polymer know-how for making the molds, which work by the age-old lost wax method. Bringing down prices of the new blades enough for widespread use required, among other things, the development of precise process controls, improved alloys, solidification models, and sophisticated furnaces.

Since 1982, single-crystal turbine blades have become standard equipment in the hot parts of engines (photo). Pratt and Whitney, PCC Airfoils, Howmet, and others have now made millions of single-crystal blades and vanes. Virtually every late-model military and civilian aircraft has them.

The need for innovation continues, however. In the quest for ever-higher engine operating temperatures, airfoil engineers are working on new blade designs and vane shapes and new internal geometries that will enhance cooling rates. Ceramic-based thermal barrier coatings to push engine temperatures higher are becoming more common. Thermal barrier coatings add the challenge of understanding and controlling interfaces between the superalloy blade and the ceramic overcoat.

Single-crystal technology is also being used by the power industry, which has been developing large, stationary turbine engines for generating electrical power. In just the last few years, the efficiency of these giant engines has nearly doubled, to about 60 percent. Part of that dramatic jump came from single-crystal turbine blade technology, which is likely to increase the demand for even better, larger, and more capable turbine parts.

Source: Giamei, 1998; Maurer, 1998; NRC, 1996.

BOX 2-2
Copper Interconnects for Semiconductor Chips

For more than 30 years, electronics engineers knew they could design faster, more capable microcircuitry if they could interconnect the ever increasing numbers of transistors on chips with copper microwiring. As it turned out, aluminum proved to be easier to integrate into the chip-making process and chip operation, so aluminum, not copper, became the standard interconnect metal of the microelectronics revolution.

The allure of copper remained intact. For one thing, copper conducts electricity with about 40 percent less resistance than aluminum. In chips, that would translate into higher switching speeds, which is the key to computing power and high performance communications circuitry. Tiny copper interconnects would also be able to withstand higher current densities so that transistors could be packed closer together, another favorite way of boosting chip performance.

Despite its potential, however, copper had a killer flaw. To chips, copper behaved like a virus because it readily diffuses into silicon and prevents transistors from functioning. Even if that were not the show-stopper for copper, there was no easy way of depositing and patterning minuscule copper wires with the uniformity to ensure high chip yields and long-term chip reliability. So for 30 years aluminum had no real competition for interconnects.

The idea of copper interconnects might never have resurfaced had chipmakers not been so successful in designing and fabricating ever more powerful chips. The relentless course of miniaturization behind this success was bound to come up against aluminum's limitations in terms of resistivity and current density. Because of the need to implement improved interconnect technology, copper interconnects were included in the industry road maps in the mid-1990s and the semiconductor industry consortium, SEMITECH, conducted research to try to scale-up the technology. And in late 1997, first IBM, and then other big league players in the semiconductor industry, revealed that more than a decade of research had finally opened the way to copper interconnects. The following year, several companies retooled their fabrication lines and actually began shifting from aluminum to copper interconnects in their high-performance chips (see photo).

Copper interconnect technology came together first at IBM for several reasons. Since the 1960s, the company had been developing expertise in electroplating high-quality thin films of copper. IBM started plating copper/permalloy thin film heads for magnetic data-storage systems in 1979. Researchers developed additional expertise in handling copper from the manufacture of printed wiring boards and high-performance chip packaging. In 1986, in a major breakthrough, researchers identified a reliable barrier to prevent copper from diffusing into the nearby silicon. With good ways of patterning copper and preventing it from poisoning the chips, the prospects for copper interconnects soared. In 1989, IBM even demonstrated the use of copper interconnects, along with a new polymer-based insulator (low dielectric constant material) on a manufacturing line.

Just as the engineering momentum for copper interconnects was accelerating toward wholesale integration into the chip fabrication process, a fundamental change in course in the history of chip technology slowed things down. In the early 1990s, the semiconductor industry shifted from using so-called bipolar transistors to CMOS (complementary metal oxide on silicon) technology. CMOS had slower

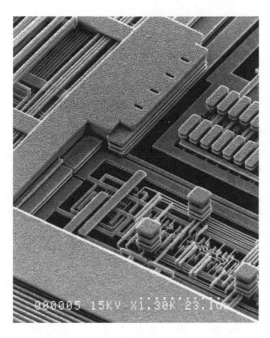

*Scanning electron micrograph of a device with IBM's first-to-market six-level cop-
per interconnect technology. Source: Courtesy of International Business Machines
Corporation. Unauthorized use not permitted.*

clock speeds but drew less power, which was becoming crucial, particularly for lap-
top computers whose utility and marketability depended on how long they could
work without having to recharge their batteries. At IBM, the shift to CMOS technology temporarily took precedence over the R&D on copper interconnects.

The program did not disappear, however. Bipolar transistors were still the mainstays of high-performance servers including those that would link into the Internet.
And as the density of CMOS increased, copper inevitably became more attractive
there as well. As the pace of research at IBM picked up dramatically, a multidisciplinary group rapidly worked out research, development, and manufacturing
details that led to the company's 1997 announcement of next-generation semiconductor chips with copper interconnects.

All the signs of an industry-wide conversion are showing themselves. Most
major semiconductor companies have announced their own goals and milestones
for implementing copper interconnect technology. Universities are offering short
courses and seminars in copper interconnect technology. Technical conferences
are including symposia on the topic. The biggest hurdle for semiconductor makers
will be investing millions of dollars for new capital equipment specially designed for
getting the best out of copper while keeping its well-known tendency to poison
electronic devices under control.

BOX 2-3
Titanium Aluminides: Unrequited R&D

Titanium aluminide alloys have been on the minds of turbine engine designers for more than 40 years. These alloys are about half the weight of nickel-based superalloys, yet can function at temperatures nearly as high. An engine made with titanium aluminide alloys could be dramatically lighter, which would make planes powered by these engines more capable.

The quest to develop titanium aluminide alloys (TiAl or Ti_3Al) began in 1956 when General Electric (GE), a maker of turbine engines, began funding studies at the Armour Research Foundation. Despite some promising results, funding cutbacks by GE forced the work to stop. Several other research groups picked up the baton in the 1960s, but when a group at Battelle tried forging these alloys with standard hammer forge equipment, the alloys shattered into bits. Word got around, and since then engine designers have associated titanium aluminides with hopeless brittleness.

A small core of researchers continued doing fundamental studies on the materials, but work toward more practical ends was not resumed until 1972, partly by accident. Harry Lipsitt and colleagues at the Air Force Materials Laboratory had been looking for materials that could meld metallic and ceramic features. and intermetallic titanium aluminides, which shatter like ceramic when struck, seemed like a good posibility. At first, the Air Force interests were scientific, but they soon became technological as well. Within two years, Air Force researchers had developed a far more ductile, workable titanium aluminide alloy. Their success shook free some funding, which was awarded to several engine makers, including GE and Pratt and Whitney. Development of titanium aluminide had still not progressed substantially, however, partly because the work was applicable to defense systems, resulting in findings and data that included proprietary information whose circulation was extremely limited.

Most research was aimed toward a dead end in terms of scaling up the technology. Then, at a technical conference in 1975, the Air Force researchers received an unsolicited tip by a vice president of Timat, a titanium metal firm. At the time, the researchers had been pursuing a powder-metallurgy approach in which titanium aluminide parts would be made from powdered starting materials shaped and baked into final shapes much as ceramic items are. But at the conference, the group learned that the titanium industry was not geared for powder metallurgy approaches, and that casting using molds and molten alloy was the way to go.

Based on this painful bit of industrial intelligence, the Air Force group shifted toward casting methods, which, although averting a dead end, set the development clock back considerably. It took another decade—until the late 1980s—before researchers had developed alloys they could work into various engine parts.

During all this time, the reputation of titanium aluminides as a brittle material persisted. Lipsitt recalls carrying lots of intricately shaped titanium aluminide parts in his briefcase to prove to people that these alloys indeed could be formed into usable components.

Today—more than 40 years after research began—titanium aluminide alloys are technology-ready. Several years ago, GE ran titanium aluminide engine parts through 1,500 cycles (heat, run, cool, repeat) with no problems. Nevertheless,

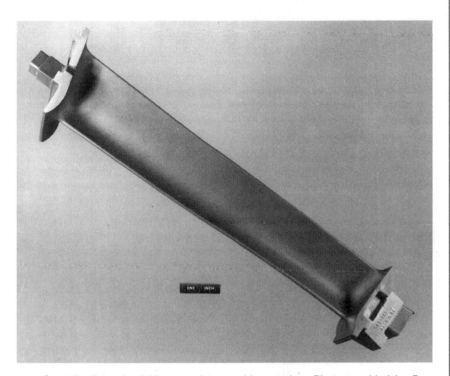

Cast titanium aluminide vane for a turbine engine. Photo provided by Dr. Harry Lipsitt.

engine designers have not yet embraced titanium aluminide. For one thing, they have a powerful incentive to not use new materials. Because the engines they build may someday carry 300 passengers at 30,000 feet, they are more comfortable working with familiar materials with long track records than with new materials that have no service track record.

Ironically, the automotive industry may be the first to adopt titanium aluminide into actual service. If fuel efficiency standards go up—as they almost certainly will—car makers will have to build smaller engines that can operate at higher RPMs. Maintaining these engine speeds will be easier with lighter weight valves that have lower inertia, which will enable them to open and close faster. Major car companies have already tested valves made of titanium aluminides for this purpose, but the alloy's cost remains a barrier. The stroke of a legislative pen calling for more efficient cars could ultimately convert more than 40 years of research on titanium aluminides into moving metal.

Sources: Personal communications between I. Amato with Harry Lipsitt, Wright State University, 1998; Allison, 1998.

BOX- 2-4
Capitalizing on Luck
The Development of Tungsten Filament Wire

The first generation of incandescent lamps a century ago had carbon filaments that were fragile, brittle, and short lived. And even while they were shining, flaws dimmed the light. Hot carbon from the filament often reacted with residual gas molecules inside the bulbs leading to black deposits on the glass jackets. What's more, watt for watt, carbon emitted fewer lumens than many other materials when electrified to incandescence.

Among the early competitors of carbon at the turn of the century, tungsten was the most promising for filaments. For one thing, its light output was three times that of carbon. Tungsten's extremely high melting temperature and the tendency of its atoms to stay put even when hot also seemed promising for building better, longer-lasting bulbs. The trouble was, no one had ever produced tungsten metal that was ductile enough to pull into a fine filament that would last, let alone a fine coil (to increase its light-emitting surface area within the bulb). Not until 1909, that is, when the tenacity, and luck, of William Coolidge of the General Electric company combined to usher tungsten in as a critical material for one the most important technologies of the modern era—incandescent lighting.

Coolidge had been working on the tungsten filament problem for three years when he finally succeeded in making short lengths of thin tungsten wire by heating square ingots of the metal and pulling them through a succession of ever-smaller wire-drawing dies, all while keeping the metal hot. During these experiments, he discovered that the very process of deforming the ingot into thin wire had some-how made the tungsten ductile. He could bend the wire cold and it wouldn't break. Coolidge didn't understand what had happened to tungsten's hidden anatomy to make this possible, but it was a pivotal advance in the history of lighting. The following year he wrote in his laboratory notebook that he was reeling long lengths of tungsten wire onto spools. The new ductile tungsten wire rapidly became the stuff of countless incandescent lamp filaments, where it is still used today.

Yet all of Coolidge's tenacity may not have paid off (at least not so soon) had he not happened to use so-called "Battersea-type" clay crucibles in the process of reducing tungsten oxide to tungsten metal powder. He had noticed that tung-sten filament made from metal produced with these crucibles lasted longer than filaments made with metal prepared otherwise. Unseen and unknown to Coolidge, potassium from the clay crucibles had leached into the powder during the reduction process and caused the structural changes that led to the fila-ment's ductility.

Under a microscope, researchers could see one important change. In materi-al produced without the Battersea-type crucible, the grains of the tungsten fila-ment aligned along the axis of the filament creating a bamboo-like structure. At high temperature, the boundaries between the "bamboo" segments would weak-en, the filaments would break, and the light would go dark. With the Battersea-type crucible, however, the grains elongated and interlocked into a much stron-ger, longer lasting, ductile architecture. Coolidge's invention of ductile tungsten gave General Electric a dominant position in the incandescent lamp filament business for many years.

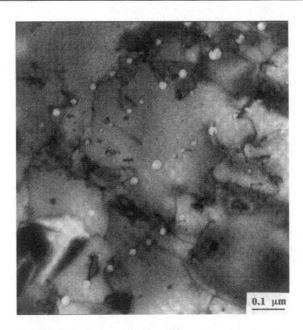

0.1 μm

Rows of potassium bubbles in annealed tungsten wire. Source: Briant and Bewlay, 1995.

Using electron microscopy, researchers had detected either bubbles or particles that appeared to control the movement of the grain boundaries and caused the interlocking grain structure. With new surface chemical-analysis techniques in the 1960s and 1970s, a team of researchers led by Westinghouse identified potassium as the trace element responsible for the formation of the bubbles, and the mystery of the role of the Battersea-type crucible was solved. As it turns out, the potassium forms gas bubbles within the metal during high-temperature processing, which elongate into tubes during the wire-drawing process. After annealing, the tubes pinch off into series of smaller bubbles often aligning in rows (photo). The micromechanical consequence is that the bubbles pin the metal's grain boundaries in place, thereby blocking the kinds of intergranular motions that lead to filament failure. Coolidge did all of that without knowing it, and his Battersea crucibles deserve half the credit.

Modern tungsten wire filaments are variations on the original theme, which is testimony to the lasting influence of Coolidge's breakthrough. Theoretical and laboratory work since the 1960s spelling out the physics and chemistry behind the original discovery have led to more deliberate methods of producing long-lasting tungsten filaments. But every time someone in the world flicks on an incandescent light, the power and payoff of a well exploited accident show up literally like a light bulb.

Source: Briant and Bewlay, 1995.

case, the technique for manufacturing superior filaments was developed long before the basic scientific underpinnings for the improvement were understood. Thus, many conceptual schemas can be generated for material/process innovations, but there will always be more exceptions than adherents for specific schema.

To ensure that all of the linkages among the MS&E and end-user communities and all of the opportunities for strengthening them were identified, the committee decided to use the simplest and broadest view of material/process development and commercialization. The conceptual schema initially adopted by the committee (Figure 2-2) was first introduced in the report *Commercialization of New Materials for a Global Economy* (NRC, 1993). The committee then modified the Activity section to emphasize that material/process development and commercialization are not linear progressions through time and that feedback could occur between any of the phases in the overall process. The committee's revised version of this conceptual schema is presented in Figure 2-3.

Of the five phases in Figure 2-3, the committee decided to concentrate on Phases 1, 2, and 3. The committee's rationale for this decision was (1) although a great deal of exploratory research might not be directed toward the development of a new material/process *per se* during the development of the knowledge base at the early end of the schema (Phase 0), this research could nonetheless yield discoveries that lead to new ideas, and (2) significant refinement of new materials/processes takes place during production and preproduction at the later end (Phase 4), and many years may pass before a profit is realized from a new process or material. Therefore, the following discussion of the various phases of the conceptual schema focuses mainly on the period between (1) the recognition and publication of a new material/process or application and (2) sign-off by an end-user on plans to put the innovation into production. Subsequent applications of the material/process technology in other products or models would be represented in the conceptual schema as a separate development, generally entering the process at Phase 2 or Phase 3, depending on the similarity of the new application and the initial application.

PHASE 0: KNOWLEDGE-BASE RESEARCH

Knowledge-base research is traditional, "why-motivated," basic research to increase the fundamental MS&E knowledge base. Ideas for Phase 0 research originate in many ways and from many sources. Researchers may choose simplified versions of problems in order to model and elucidate basic principles. Ideas may also be derived from further up the materials-development process when problems are encountered that require a more thorough understanding of the fundamental behavior of materials for their resolution. Phase 0 research is predominantly conducted in university and government laboratories, and the results are usually presented at academically oriented conferences or published in peer-reviewed journals and Ph.D. dissertations. Basic-research-oriented federal, state,

ACTIVITY	Phase 0: Knowledge-Base Research	Phase 1: Material Concept Development	Phase 2: Material Process Development		Phase 3: Product Integration	Phase 4: Product Development
INDUSTRY TERMINOLOGY	Technology-Base Development		Product Development and Demonstration		Early Commercialization	Full Commercialization
DOD TERMINOLOGY	6.1 Basic Research	6.2 Exploratory Development	6.3 Advanced Development		6.4 and 6.5 Engineering Design/Operational Systems Development	Initial Operational Capability; Operation and Maintenance
			6.3A Component Development	6.3B Feasibility Demonstration		

FIGURE 2-2 Conceptual schema of materials development and commercialization processes. Source: NRC, 1993.

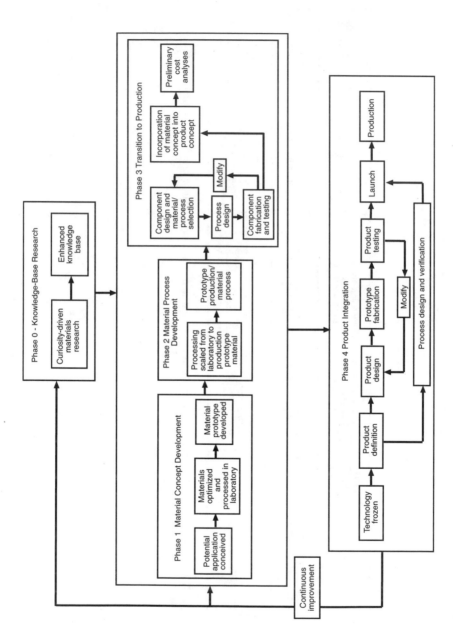

FIGURE 2-3 Revised conceptual schema of materials/processes development and commercialization.

or private foundations (e.g., the National Science Foundation) generally fund Phase 0 research. Linkages to these institutions, therefore, is critical to the success of any R&D program, both for funding and because of an increase in the number of multidisciplinary projects that require expertise in a variety of fields. The linkage between Phase 0 researchers and end-users is also important, both as a source of ideas for researchers and a source of job opportunities for graduates.

Although it is very difficult to assess the efficacy of exploratory research in terms of the investment required to yield discoveries of a particular value, other issues should be considered in assessing the value of knowledge-base research. Phase 0 research by single investigators and multidisciplinary teams has been very productive in the United States, as shown by the large number of Nobel Prize winners and new ideas for materials/processes that have been developed. The value of Phase 0 research, however, goes beyond the discovery and initial development of new materials/processes. First, basic research is the educational basis of the next generation of scientists and engineers for academia and industry. Second, basic research increases the knowledge base on which the continued evolution and incremental improvements of current materials are based, as well as the discovery of leapfrog technologies that dramatically increase industrial competitiveness. Thus, even though some areas of basic research never lead to the application of new materials/processes into final products, they are still of great value to the MS&E community. A summary of Phase 0 research is presented in Figure 2-4.

PHASE 1: MATERIAL CONCEPT DEVELOPMENT

Phase 1 concept development begins with "use-motivated" research, either as (1) the next logical step in the sequential development of an idea from Phase 0 (Box 2-5) or (2) from a serendipitous discovery of a promising material/process phenomenon. Phase 1 research includes more detailed investigations of materials/processes to determine their significant properties and parameters and their true merits. Phase 1 research is conducted at universities, government laboratories, and industry laboratories and is funded by basic-research-oriented federal and state agencies, private foundations, consortia, venture capitalists, and the industries themselves.

During Phase 1, researchers or industries usually apply for the patents for their concepts, either to protect their concepts during later development or to ensure that the researchers or companies retain the right to use the materials/processes even if another company eventually develops them. Potential leapfrog or high-impact innovations, however, may be patented as early as Phase 0. Once patent protection has been obtained, the early results may be either presented at academically oriented conferences or published in peer-reviewed technical journals or dissertations.

Mission-oriented federal or state funding agencies (e.g., the U.S. Department

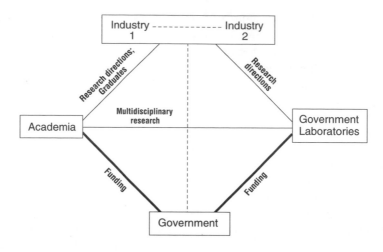

Intent	Basic research (can be based on feedback from later stages)
Starting point	Existing knowledge
Result	Enhanced knowledge base
Output	Peer-reviewed publications; conference proceedings; disserta tions; graduates
Principals	Predominantly university and government research laboratories; some industrial laboratories
Funders	Predominantly federal/state government; some industry/consortia
Gatekeepers	Predominantly federal/state government; some industry/consortia; research peers
Time period	5 years to more than 20 years

FIGURE 2-4   Characteristics of knowledge-base research (Phase 0). The thickness of the line indicates the importance of the linkage.

of Defense or U.S. Department of Energy) may fund three- to five-year follow-up projects in the later stages of Phase 1. This work is still performed largely at university, government, or industrial research laboratories but focuses on how small variations in processing affect subtle (and difficult to measure) properties. Parallel attempts might be made to improve processing equipment and procedures. Toward the end of Phase 1, researchers often perform simple cost analyses based on laboratory processing to demonstrate the potential economic advantages of the new material/process over existing technologies.

Finding 2-4. Only very rudimentary cost analyses can, or should, be attempted during Phase 1. At this stage, there are too many variables (e.g., other technological advances, government regulations) to model costs accurately. Inconclusive or negative results from cost modeling could inhibit research that could still have great potential.

Researchers often present the later results from Phase 1 research at industry conferences and publish them in more applied, peer-reviewed journals or as masters theses. Phase 1 ends with a proof of concept and preliminary quantification of the technical risk that can be used to attract initial industrial interest. In many cases, however, companies that might use the new material/process are not aware of assessments of technical risks and benefits, and companies that are aware of them may not be convinced of the potential applicability, economic advantages, or scale-up potential of the innovation.

The linkages between Phase 1 researchers and potential federal and state sponsors of a study are critical. The linkages between researchers and research institutions are also extremely important, especially for multidisciplinary projects that require expertise in a variety of fields. Linkages with end-users and venture capitalists become even more important during Phase 1 because their support will determine whether a material/process is developed further. A summary of Phase 1 research is presented in Figure 2-5.

Finding 2-5. Although linkages among researchers, potential federal and state sponsors, and research institutions are critical during Phase 1, linkages with end users and other business or financial interests become increasingly important. Their interest and support will determine if the ultimate potential of a project is realized.

PHASE 2: MATERIAL/PROCESS DEVELOPMENT

Phase 2 represents a wide variety of activities and periods of inertia between Phase 1 research and Phase 3 product integration. Phase 2 is extremely important, however, because it is the transition point between "technology push" from the MS&E research community to "product pull" from the end-user community.

The primary objective of Phase 2 R&D is to scale-up production from the laboratory to the prototype level to quantify the business risks in the innovation. Depending on the industry, prototype production can range from proving that the material/process can be integrated into existing production methods to the fabrication of a pilot plant. Phase 2 R&D ends when the quantification of the business risk shows that the innovation has both scale-up potential and economic advantages, and industry decides to integrate the technology into a product. Phase 2 R&D can last as long as 20 years or can be skipped over entirely for particularly promising or simple-to-implement changes (Box 2-6).

Despite the potential for social or economic rewards offered by many material/process innovations, Phase 2 has been referred to in the MS&E community as the "valley of death" because mechanisms to encourage Phase 2 activity

BOX 2-5
Great Good Fortune: Data Storage, Magnetoresistance, and Giant Magnetoresistance

The hunger for more data storage has become almost as certain as death and taxes. Everyone in the data-storage business knows that even the high-end data-storage technologies of the moment will, sooner or later, fail to satisfy that hunger. It's an innovate-or-vanish arena.

The only constant in the half-century old Computer Age has been the dominance of magnetic materials for the mass storage of data. The scheme is simple: partition the magnetic material into areas, called bits, and let two possible directions of magnetization of each bit represent the ones and zeros of digital data. Assigning the ones and zeros to the bits has been a matter of applying a magnetic field strong enough to induce the desired magnetization; reading the data has been a matter of detecting the magnetizations and converting them into electrical signals the rest of the computer can process.

Throughout the 1960s and 1970s, the so-called inductive read/write head had been doing just fine in data-storage systems. When an electric current ran through the coil, it created a magnetic field that could write the underlying bits on a tape or disk. When the head flew over a bit, an electrical voltage was induced in the coil, transferring the stored data into the computer via an electric signal. For years, engineers were able to cram more and more data onto disks and tapes by, among other things, shrinking the area of each bit. But in the 1980s, the bit size approached its limit, beyond which it would be too small—and the magnetic fields too weak—for the inductive heads to read accurately.

As early as 1970, it had been suggested that a class of magnetoresistive (MR) materials, whose electrical resistance would change a few percent depending on the direction of magnetization, could be used. With sensitive MR materials, such as the nickel-iron metal known as Permalloy, engineers could continue to shrink bit sizes. The details of manufacturing MR materials into heads became a high priority for IBM. Competition in data storage had become fierce, and any dramatic new technological advance would be valuable. In 1987, the company introduced the first commercial MR read heads in a tape-based storage system; three years later they were introduced in hard-disk drives. The use of MR materials in data storage by a major company like IBM helped accelerate the rate of increased storage capacity to 60 percent per year.

The insatiable hunger for storage capacity that led to MR heads eventually outstripped the capacity of practical MR materials. In the late 1980s, reports by French and German researchers of materials with giant magnetoresistive (GMR) effects tantalized the data-storage industry. The resistance of some of these very thin multilayered structures changed by more than 100 percent making them tens of times more sensitive than the MR heads then just becoming part of data-storage systems. These materials triggered a worldwide competitive scramble in both basic and applied research for practical materials whose GMR effect would be large even at room temperature and small magnetic fields. One group at IBM made and tested about 30,000 different multilayer combinations using different elements and different layer thicknesses.

Images of data bits on a magnetic hard disk as seen by a magnetic force microscope. The data-bit density at the far left is equivalent to 10 billion bits per square inch when adjacent tracks are included. Photo by Tom Chang, IBM Storage Systems Division. Courtesy: International Business Machines Corporation, Almaden Research Center. Unauthorized use prohibited.

The promise of GMR materials for boosting data storage capacity was so great that eight companies and six universities joined in 1992 into an intense precompetitive collaboration through the U.S. Department of Commerce's Advanced Technology Program, in this case administered by the National Storage Industry Consortium. The five-year effort yielded critical insights and engineering knowledge at a pace that would have been unlikely if the participants had been working separately. In early 1998, IBM delivered the first disk products in the world using GMR heads. The company proudly advertised hard-disk drives for desktop computers capable of storing 16.8 billion bytes of data (see photo), about 20 times the capacity possible less than five years ago.

Although multibillion-byte hard drives are impressive now, the hunger for more storage capacity will continue. So the same basic motivation that drove data-storage engineers to embrace MR materials and then GMR materials will eventually force them to abandon magnetic storage techniques altogether. Perhaps miniaturized atomic-force microscopes will one day be reading and writing a trillion bits of data, each one embodied by a cluster of atoms, all of them residing on a square inch of some yet to be identified storage medium.

Sources: Almasi et al., 1972; Baibich et al., 1988; Dieny et al., 1991; Egelhoff et al., 1996; Hunt, 1970.

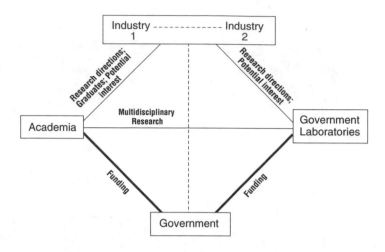

Intent	Proof of concept (i.e., properties and characteristics defined; materials optimized)
Starting point	Potential end-use application conceived
Result	Material/processing prototype; laboratory-scale processing (some processing proof); quantification of technical risk
Output	Patent(s); technical and peer-reviewed publications; Ph.D. and masters theses; graduates
Principals	Predominantly university and government research laboratories; some industrial laboratories
Funders	Predominantly federal/state government; some private indus try; entrepreneurs
Gatekeepers	Predominantly federal/state government; some industry/consortia; research peers
Time period	3 to 10 years

FIGURE 2-5   Characteristics of material concept development (Phase 1). The thickness of the line indicates the importance of the linkage.

are inadequate (Burte 1981). Recently, however, federal and state governments and consortia have shown more interest in Phase 2 R&D (see Chapter 3). For example, programs such as the University-Industry Program of the New York State Science and Technology Foundation are providing matching grants for the establishment of university incubators to facilitate the implementation of new technologies in state industries, improving both companies' competitiveness and the state tax base. A summary of Phase 2 R&D is presented in Figure 2-6.

Finding 2-6. The importance of Phase 2 R&D and the substantial differences between Phase 2 and the traditional Phases 0 and 1 research are gaining recognition with funding agencies, universities, government laboratories, and industry. Overcoming the barriers to Phase 2 R&D is the most promising way to shorten the time to market of laboratory innovations.

Based on the information provided at the workshops, the committee identified six principal barriers to the performance of Phase 2 R&D: variability and instability in funding; the high costs and long time frames associated with certification of materials/process technologies; the difficulty of accurately modeling implementation costs and demands for materials; the multidisciplinary nature of the R&D; the difficulty of mobilizing academic research; and the differences in end-user and research cycle times. These barriers are discussed in the following sections.

Funding Gap

The first barrier to Phase 2 R&D is the perception that funding is variable and unstable. The funding bases for Phases 0, 1, 3, and 4 are relatively stable and efficient. Phases 0 and 1, which are nationally and internationally motivated by the desire to improve industry, society, and the human condition (see Chapter 1), are supported by federal and state government programs and encouraged at universities and government laboratories in a variety of ways (e.g., tenure, peer-reviewed publication, and research awards). Phases 3 and 4 are a natural part of doing business and are motivated by industry's desire to remain competitive. For most (if not all) industries, survival depends on how well a company interprets market forces and implements new technologies in response.

Phase 2 R&D, however, requires taking tremendous risks and can be the most expensive research phase, especially if the construction of a pilot plant is involved. Phase 2 R&D can also have the highest payoff, however. Industry generally prefers that many different Phase 2 programs be under way at any given time to increase the chances of finding new and potentially profitable technologies and to increase its options for meeting new economic or environmental requirements. Responsibility for the funding of Phase 2 R&D is a matter of debate, however. Many in industry believe that funding for Phase 2 R&D should be the responsibility of the federal government because it enhances national economic competitiveness. Many federal policy makers believe that industries should be responsible for funding their own Phase 2 R&D because it is in their competitive interest to do so. Because of the lack of secure funding for Phase 2 R&D, universities rarely have either the wherewithal or access to state-of-the-art industrial equipment to participate. In many cases, the MS&E R&D community attempts to attach Phase 2 R&D to Phase 1 research programs and adapt existing equipment as best they can.

BOX 2-6
Accelerated Innovation in the Semiconductor Industry

New, more capable materials are the flesh and bones of new, more capable products. For some industries, including the semiconductor industry, a promising new material or process can be used in products in a matter of months. In more conservative industries, the integration of a new material can take decades. Titanium aluminides—lighter weight, higher temperature alloys for turbine engines—have been in development since the mid-1970s and are still not the stuff of airplanes. The vast differences in the rates of integration of new materials into products by different industries reflect a complex web of factors, including technical details of the technologies themselves, the cultures and histories of the industries, and even macroeconomic factors like the cost of commodities.

The semiconductor industry has become famous for the fast pace of its materials development and commercialization cycles. In the early 1970s, Intel co-founder Gordon Moore argued that microchips would double in computational power and halve in price every 18 months. This became known as "Moore's Law," and living up to it has been both a cause and an effect of the insatiable demand for cheaper, more powerful computing power. Moore's Law also created a competitive context in which companies must be able to make rapid incremental improvements to integrated circuits and other computer components. To do this, the industry as a whole has had to compress what this report has identified as Phase 2 and Phase 3 of the materials development and commercialization cycle, the R&D phases that bring a material with proven laboratory promise for some technological purpose through the expensive and risky work of proving its worth for integration into products.

In the semiconductor industry, many factors have come together to speed up the materials development and commercialization cycle. For one thing, its products, including personal computers, are modular, so material or process innovations can be easily implemented in different components of the final products. The limited liability of flawed technology compared to the liabilities in the automotive and aerospace industries has also accelerated R&D in the semiconductor industry. Although computer crashes are costly and troublesome, they are never by

---

Finding 2-7. Linkages among the academic MS&E R&D community, industry, federal and state funding agencies, and entrepreneurs are generally weak, and there is no consensus as to who should be responsible for the identification and funding of Phase 2 R&D programs.

Materials/Process Certification

The high costs and long time frames associated with certifying a material/process innovation are a second barrier to Phase 2 R&D. The time from innova-

themselves fatal. Materials and processes for semiconductors have been studied extensively, so the seed corn of innovation—the knowledge base—is present in abundance.

There also are a handful of sociological and economic factors that enable the semiconductor industry to compress Phases 2 and 3 of the materials development and commercialization cycle:

1. It's a relatively young industry, on the early part of the "S"-shaped curve that depicts the development cycle of many industries, when rapid development is most likely.
2. Society's wholesale adoption of semiconductor technology has enabled the industry to build and upgrade its physical and talent infrastructure rapidly.
3. To meet demand at the rate suggested by Moore's Law, separate companies have had to overcome the go-it-alone mentality of previous technological eras. For example, many semiconductor companies have pooled their resources into meta-organizations (such as SEMI/SEMATECH and NEMI) to push through the expensive, high-risk Phase 2 of the R&D cycle.
4. These companies, along with academic and government organizations, have also formulated technology road maps charting out long-term strategic goals for the industry and identifying the most-likely-to-succeed tactics for meeting those goals. The road maps have helped companies steer finite funds and resources in the directions most likely to pay off.
5. The intimate connection between basic science and technology in the semi-conductor industry has attracted the attention and talent of universities. Some academic institutions have even set up centers dedicated to research that feeds into both Phase 2 R&D and adds to the fundamental knowledge base.

The situation may change in the early decades of the next century when micro-electronics components will be so small that quantum effects will overtake classi-cal electronic behavior. At that point, the hard-won knowledge base of classical electronics and conventional lithographic-based chip-making will no longer feed smoothly into Phase 2 and 3 R&D. But, there may be a consolation for the post-Moore's Law electronics industry. The first personal quantum computer might take a lot longer to become obsolete than any 20th century PC.

tion to implementation depends on the application and generally increases with the complexity and potential liability of the application (e.g., from sporting goods to electronics to automobiles to aerospace systems). From the perspec-tive of individual companies, the barrier to supporting Phase 2 research is most difficult to overcome for technologically mature or high-liability industries because the introduction of new materials/processes requires extensive—and expensive—product recertification. The time required to certify a new material/process often approaches the limits of the patent-protection period; thus a com-pany may not have time to recoup its R&D investment before its competitors can legally use the technology. Thus, the high costs and long times associated

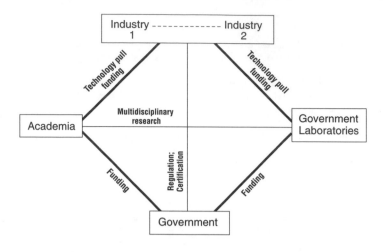

Intent	Initial scale-up from laboratory-scale process to prototype production
Starting point	Materials/processing prototype; laboratory-scale processing
Result	Prototype production (integration proof or pilot plant, depending on industry); quantification of business risk
Output	Internal industry reports; consortium sharing of database information
Principals	Industrial R&D laboratories; industrial consortia; universities as subcontractors
Funders	Predominantly federal/state government; in-kind and small amount of financial support by industry, usually via consortia; entrepreneurs
Gatekeepers	Federal/state government; private industry via consortia
Time period	0 to more than 20 years

FIGURE 2-6   Characteristics of material process development (Phase 2). The thickness of the line indicates the importance of the linkage.

with recertification tend to bias industry toward incremental improvements in developed technologies that can be implemented quickly and that allow them more time during the patent-protection period to accrue profits and recoup R&D investments.

Finding 2-8. The extended period of time and significant investment required to certify new materials/processes in technologically mature or high-liability industries are impediments to material/process innovations, especially when the time to certification and first application can exceed the patent-protection period and limit the company's ability to recover R&D investments. Linkages between industry and government regulators are important for determining whether investing the time and financial resources required to certify a new material is worthwhile in a given situation.

Modeling of Implementation Costs and Materials Demand

The third barrier to the success of Phase 2 R&D is the difficulty of accurately estimating the costs, trade-offs, and eventual demand for a materials insertion at this early stage in the implementation process. Nevertheless, because business risks must be quantified during Phase 2, better methods for modeling costs and trade-offs could demonstrate the true potential of new materials/processes.

Finding 2-9. The risks associated with a materials insertion cannot be quantified accurately with current methods of estimating costs and eventual demand and for making trade-offs. Modeling methods must be improved to assess the true potential of new materials/processes.

Multidisciplinary Nature of Phase 2 R&D

The fourth barrier to Phase 2 R&D is the wide spectrum of expertise required to complete the development of materials/processes. The required expertise is generally beyond the ability of any one individual and could require the formation of multidisciplinary teams. Linkages among universities, government research laboratories, and industry are thus important for amassing the required expertise. Some potential methods for promoting multidisciplinary research projects within and among universities, government laboratories, and industry and for promoting interaction are (1) permitting Ph.D. and masters students to conduct research in industry; (2) promoting short-term exchanges (e.g., one-day consultancies to one-month visiting positions), as well as long-term sabbaticals among universities, government laboratories, and industry; and (3) encouraging industry researchers to seek adjunct positions at local universities and government laboratories. Researchers will require management and interpersonal skills to function well in multidisciplinary teams.

Finding 2-10. Phase 2 R&D is becoming increasingly multidisciplinary and dependent on (1) the promotion of multidisciplinary research projects within and among universities, government laboratories, and industry and (2) the availability of people trained to work on multidisciplinary teams.

Mobilizing Academic Research

Another barrier to Phase 2 R&D is the difficulty of mobilizing academic researchers to perform Phase 2 research. The driving forces for academic and end-user communities differ, and an academic R&D program must not only have funding and equipment but must also fit into the academic culture and educational mission of the university. For example, one problem commonly encountered by universities is difficulty evaluating junior faculty members engaged in Phase 2 R&D. Tenure appointments are generally based on the publications record of the candidate and assessments by recognized faculty members at other institutions. Limiting the opportunities for junior faculty members to publish the results of their research in the open literature or limiting their ability to collaborate with other faculty members places them at a disadvantage when being considered for tenure.

Finding 2-11. Phase 2 R&D, which involves interacting with industry and other nonacademic organizations, is often hindered at universities because the traditional methods of evaluating research faculty for tenure do not value participation in Phase 2 research projects as highly as Phases 0 and 1 projects.

Product Cycle Times

Differences between academic and industrial product cycle times can also cause problems. All student research must have major teaching and educational components that must be conducted and published within designated time periods (i.e., B.A./B.S. degrees in four years; M.A. degrees in one or two years; Ph.D. degrees in five or more years). Industry has very different funding and planning cycles, however, and can rarely plan or fund more than a year in advance. As a result, much of the industrial research conducted by academic institutions is short term (i.e., one year or less).

Finding 2-12. Industry's funding and planning cycles tend to be incompatible with the time frames and commitments required for educating graduate-level students.

PHASE 3: TRANSITION TO PRODUCTION

Phase 3 is the best defined of the R&D and commercialization phases but varies the most from company to company and from industry to industry. Phase 3 begins when a company becomes convinced of the cost/benefit advantages of a new material/process and schedules it for integration into a final product. The objective of Phase 3 R&D is to make the transition to reliable, full-scale production without compromising the advantages of the materials/process innovation. In addition to business and marketing issues, six major technical issues must be considered:

- Can the materials/processes be optimized during scale-up?
- Are reliable sources for the materials/processes available?

- Can existing equipment be adapted to produce parts to specification using the new materials/process, or is a new infrastructure required?
- Can quality be adequately controlled for the materials/process innovation and resulting products?
- Are techniques available for integrating the parts produced with the materials/process innovation?
- Is extensive training required to implement the new methods/technologies?

Traditionally, Phase 3 R&D has been proprietary and was conducted predominantly by industry—generally by the original equipment manufacturer (OEM) and one or more suppliers. Thus, the critical linkages are usually either between the industrial R&D and manufacturing branches of a single company or between members of the supply chain of an industry. Occasionally, universities or government laboratories are contracted, and sometimes companies attempt to develop a material/process jointly.

There are several routes by which a "technology pull" may be established and a technology in Phase 2 (or even at the end of Phase 1 for exceptional breakthroughs) acquired by a company for Phase 3 scale-up. Regardless of the route to Phase 3, an industry champion who can persuade decision makers that the technical and business risks of introducing the new technology are justified may be critical.

Development engineers might read about a new laboratory material/process in a technical journal and decide that it holds sufficient promise for investment. The company might then contact the original researchers and offer to work with them on a proprietary basis. The company might also try to glean whatever information it can from published results and discussions with the researchers and then initiate its own program. Searches of the open literature are especially effective if a major new driver (e.g., a new federal regulation) is introduced and an industry must respond quickly to remain competitive. Linkages between the end-user community and university or government research laboratories must be strong for a new technology to make the transition in this way.

An innovation may also come to the attention of a company via a supplier or competitor that has focused on a new material/process. This often happens in the United States and might be the dominant driver for entering Phase 3. In many industries, notably the automotive sector, competitors' products are routinely disassembled and examined for new engineering approaches, manufacturing techniques, and materials (see Box 2-7). A related mechanism is for an innovation developed for one application, often through government-sponsored programs, to be adapted for application in another industry. This allows manufacturers to take advantage of work performed by others to improve products and provides the materials supplier an opportunity to recoup a portion of the development costs. An example of this mechanism is the use of advanced structural materials in consumer sporting equipment (e.g., carbon composite rackets and golf club shafts

BOX 2-7
Tailor-Welded Blanks

Recognizing good ideas first developed by others is a great way of shortening the materials R&D process. The development of so-called tailor-welded blanks in the U.S. automobile industry is a case in point. Traditionally, structural auto body parts have been made by cutting steel sheets (with a specific thickness, coating, and set of metallurgical properties chosen for the application) into specific starting shapes called blanks. The blanks are then stamped into the three dimensional forms of finished parts. Many assemblies—including body side panels, wheel housings, and fenders—require that some areas be reinforced with heavier steel for safety or to withstand stresses. For decades, carmakers have made these heterogeneous assemblies by first making individual parts—individually designed and formed to have the needed properties—and then welding these parts together into finished assemblies.

In 1990, when U.S. engineers disassembled and examined the new models of the Lexus LS400 automobile, they were hoping to find an innovation that would simplify carmaking. Underneath the LS400's sleek exterior, the engineers found a variety of applications of what later became known in the industry as tailor-welded blanks (TWBs). The key innovation of these blanks is the incorporation of the heterogeneous material properties needed for car components into a single blank that can be formed into the finished component with a single set of dies. The blanks are tailor-made by laser-welding together flat steel sheets with different strengths, thicknesses, and coatings. European and Japanese carmakers were the first to use TWBs during the 1980s. Their use in the 1990 Lexus LS400 was an effective wake-up call to the U.S. car industry. Within only three years, an entire industry to supply TWBs to manufacturers had taken root and begun to grow.

Picking up on someone else's good idea shaved at least a decade off the normal time for a new material or materials process to wend its way into service from the time of the original innovation. One reason for the rapid development of a TWB supply line in the United States was that the technology was related to already mature laser-based welding processes (photo). Much of the development work was done by agile start-up companies with laser welding expertise, such as LaserFlex and Utilase. Another key to the rapid development of TWBs was the creation in 1992 of the TWB Company—a joint venture between Worthington Steel, a major intermediate steel processor based in Columbus, Ohio, and German-based Thyssen Krupp-Stahl AG, which had pioneered TWB-technology in the early 1980s. In 1997, several major steel companies joined the company as limited partners, accelerating further diffusion of the technology through the entire manufacturing chain.

The use of TWBs by U.S. carmakers has been growing steadily (photo). According to a report by the American Iron and Steel Institute on efforts to develop ultralight steel auto bodies, TWBs are central to the steel industry's bid to remain a mainstay of the automobile industry (even though TWB technology could also be used for aluminum). Pursuing business as usual would be risky for the steel industry in the face of growing competition from nontraditional materials, including aluminum and composites. TWB technology is an important factor in this competition because it can not only simplify and improve the manufacturing process, it can

Automated welding system for the production of door inner blanks. Source: TWB Company.

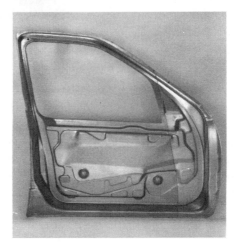

Laser-welded door inner panel. Source: TWB Company.

also save money by reducing the amount of steel needed to make vehicles as well as reducing scrap volumes.

Just how far TWBs infiltrate the U.S. auto industry will depend on the suppliers' cost effectiveness and willingness to innovate, as well as on how far carmakers are willing to move away from conventional manufacturing practices. The progress so far is testimony to what can happen when the "not invented here" syndrome does not obscure the technological potential of someone else's good idea.

Source: McCracken, 1998.

and titanium golf club heads). Thus, linkages between industries—even if they are not official, established collaborations—are important for the creation of "technology pull."

In recent years, consortia of universities, government laboratories, and industries have focused on Phase 2 R&D and attempted to demonstrate feasibility through precompetitive programs. Participation in these consortia allows companies to observe and evaluate Phase 2 R&D. If a company is persuaded that a technology has cost/benefit advantages, it could decide to proceed with in-house development on a proprietary basis. Consortia are typically funded by the federal or state governments for the university/government component and by industry for the industrial component (the objectives and mechanics of consortia are discussed in greater depth in Chapter 3).

Phase 3 development is very efficient and is driven by the natural selection process of the marketplace. The process is also highly developed and almost inscrutably complex to those outside a company because decisions are based on a broad range of factors. For example, the implementation factors for the turbine-engine industry include market competition; customer needs; partnerships and licensing requirements; technology cost; technology maturity (e.g., manufacturability); risk (e.g., liabilities); resource allocation (e.g., capital outlay); enhancing versus enabling capabilities (i.e., technical merit); supplier base readiness and feasibility; and dual-use versus strategic-fit (Roberge 1998). A summary of Phase 3 development is presented in Figure 2-7.

PHASE 4: PRODUCT INTEGRATION

If the prerequisite knowledge and supplier bases for full-scale production can be successfully produced within the time limits demanded by the product development cycle for an industry, the material/innovation will enter Phase 4 development and be integrated into the final product. Because product development cycle times are currently being decreased, actual R&D cannot be conducted

TABLE 2-2 Characteristics of Product Development Phase (Phase 4)

Intent	Product concept to product design to production
Starting point	Product concept
End point	Product production
Output	Full scale-up (only troubleshooting R&D is performed because cycle time is short); internal configuration-management documentation
Principals	Industry; federal end users; customers
Gatekeepers	Industry management
Time period	2 to 5 years

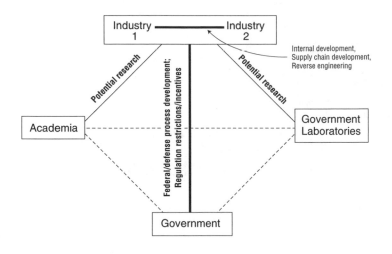

Intent	Component design and testing; materials process design and scale-up
Starting point	Prototype production
End point	Incorporation into product concept
Output	Internal company data and know-how to begin production
Principals	Predominantly industry with some researchers (university/government laboratory/industry)
Funders	Predominantly industry via internal funding; federal end-users (e.g., DOD, NASA); venture capitalists
Gatekeepers	Industry management; federal end-users
Time period	2 to 5 years

FIGURE 2-7 Characteristics of transition to production (Phase 3). The thickness of the line indicates the importance of the linkage.

during Phase 4. The MS&E R&D community still plays an important role in the successful launch of a new material/process, however, but it is usually an advisory role. The R&D community must be available in a consulting capacity to monitor field failures and advise the manufacturer on how to resolve problems. Phase 4 also provides important feedback for the R&D community on manufacturing problems or limitations that may suggest areas for further research to improve manufacturing processes. Because of the urgency and proprietary nature of Phase 4 scale-up, the in-house industrial research community is most often involved during this phase, although academic researchers are also regularly brought in on short-term, proprietary consulting contracts. A summary of Phase 4 development is presented in Table 2-2.

3
Linkages between the MS&E and End-User Communities

THIS CHAPTER PRESENTS AN ANALYSIS of the five main types of linkages between the MS&E R&D and end-user communities—industry-industry; industry-university; industry-national laboratory; industry-government; and government-research institution. University-national laboratory linkages are not discussed in this chapter because the sole reason for their interaction is to augment their multidisciplinary programs with additional expertise.

The number of interactions and collaborations that can be envisioned between the various segments of the MS&E R&D and end-user communities is nearly boundless. Nevertheless, focusing on the simplest form of each linkage can reveal specific strengths and weaknesses. Thus, this chapter will examine each type of linkage as a one-on-one interaction. The chapter will conclude with a discussion of consortia, which is the main mechanism currently used for joint ventures and interactions with participation from multiple segments of the MS&E and end-user communities.

INDUSTRY-INDUSTRY LINKAGES

Interactions among industries form the basis of all business. Since the objective of this report is to strengthen the connections among the MS&E and end-user industries, the discussion in this section focuses primarily on MS&E R&D linkages among industries.

The committee divided the typical user chain for the materials production cycle into four main sections to simplify the description of linkages between materials-based industries (Figure 3-1). The first section, *materials suppliers*, includes companies that produce the raw or semifinished materials used in the

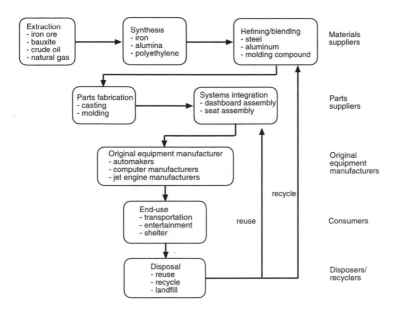

FIGURE 3-1 Typical user chain for materials production cycle, from raw material to the ultimate destiny of all materials.

fabrication of subcomponents or parts for finished products (e.g., Oremet or Carpenter Technology for the jet-engine industry; Alcoa for the automotive industry; Shipley or Ciba-Geigy for the integrated circuit fabrication industry). These companies may be involved in the extraction, synthesis, or refining/ blending processes shown in Figure 3-1. The second section, *parts suppliers*, includes companies that produce the parts used in the assembly of the final product or its subcomponents (e.g., Howmet or Ladish for the jet-engine industry; Eaton or Budd for the automotive industry; Intel or Motorola for the computer-component industry). This section is shown in the parts fabrication segment of Figure 3-1. The third section, *original equipment manufacturers*, includes both assemblers of major subcomponents (e.g., Lucas or Bendix for the jet-engine industry; Delphi or Nippondenso for the automotive industry; ReadRite or Seagate for the computer-component industry) and the main assemblers and distributors of final end-use products (e.g., GE, Pratt and Whitney, or Rolls-Royce for the jet-engine industry; Ford or Honda for the automotive industry; Compaq, Apple, or IBM for the computer industry). This section encompasses the systems integration and OEMs boxes in Figure 3-1. The fourth section, *disposers/recyclers*, includes disassemblers, recyclers, and disposers of the final products at the end of their service life (e.g., Huron Valley Steel drains, disassembles, separates, shreds, and recycles cars for the automotive industry).[1] This section encompasses the disposal box of Figure 3-1.

Materials Suppliers

Primary materials suppliers (e.g., steel, aluminum, and plastic resins) supply raw and processed materials to both parts suppliers (at all tiers) and OEMs. Generally, materials suppliers sell their products to many industries and are, therefore, not commercially dependent on any one business for their livelihoods. In the past, primary materials suppliers were only involved peripherally in the design process. As the competition for primary materials has intensified, however, they have become increasingly involved in developing their own design activities. Many materials suppliers are now being driven further up the value chain of the materials production cycle and have become involved in the OEM's product development and design processes. This is especially true for new materials concepts, for which the supplier infrastructure might not be able to meet the needs of industry or for which prospective suppliers may have underestimated the challenges of scaling up an unproven technology.

In many cases, primary materials suppliers are supplemented by specialty materials suppliers, which produce more advanced materials. Specialty materials suppliers can often be classified as "value-added distributors." For example, jet-engine alloys require specialty materials suppliers because they are a complex and carefully controlled combination of many elements combined by special processes and equipment. Although proprietary alloys are frequently developed by OEMs, specialty metals companies melt and combine the ingredients that go into a jet-engine alloy and perform a host of additional value-added activities to ensure the quality and integrity of the alloys. Similar specialty materials producers are involved in other industry supply chains, even though the supplier, not the OEM, usually develops the materials. For example, compounding companies that supply materials to the molded-plastic component industry combine constituent ingredients to create customized plastic compounds. Producing and supplying polymer compound materials for the electronics industry is a $4.0 billion business (e.g., Shipley formulates photosensitive polymers used to pattern integrated circuits, Ciba-Geigy supplies polymers used in printed wiring boards).

The sources of materials/processes innovation vary from industry to industry. For example, materials innovations in the jet-engine industry originate predominantly in the OEM's laboratory. Each innovation is considered proprietary and is a carefully guarded secret because of its potential competitive advantage,

---

[1] Similar disposal companies do not exist for the computer or jet-engine industries. OEMs in the computer industry recycle some materials, but most systems currently end up in landfills. Jet engines are too valuable to be junked entirely. Most engines are rebuilt piecemeal during repair using replacement parts. The parts suppliers usually recycle the materials from old parts.

which could translate directly into increased market share. The automotive industry, however, relies heavily on materials suppliers for materials/process innovations. These firms range in size from small, niche enterprises to very large corporations (e.g., Alcoa). In the automotive industry, suppliers market their innovations by developing ties with OEMs and parts suppliers and publicizing the potential advantages of their innovations. Materials suppliers must present material properties in terms that are relevant and understandable to designers, who are most likely to decide which materials will be used (Buch, 1998).

Recommendation 3-1. Materials suppliers should collaborate with end users to determine the type of data most useful for product designers in assessing new materials/processes and determining their suitability for incorporation into a product. Materials suppliers should be responsible for conducting performance tests to reduce the redundant materials testing by many industries.

The factors that limit the ability of the materials-supply industry as a source of innovation are similar to the problems facing parts suppliers (e.g., large capital investments, limited resources, equipment manufacturer's need for multiple suppliers). The problem is exacerbated, however, by three factors. First, the profit margins for many materials innovations are minimal, at best. The initial production volumes for advanced materials are usually limited, and alternate markets that could provide large returns on investment are rare. Thus, many potentially useful materials are not developed beyond Phase 1 because it is simply not cost effective for a materials supplier to use its limited resources to develop and market them. Second, materials suppliers for OEMs that usually develop their own materials (e.g., jet engines) must circumvent the "not-invented-here" fears latent in those industries (Maurer, 1998). The ability of end users to exploit new technologies is limited because even seemingly insignificant changes in materials (e.g., the presence of trace elements in bulk materials or a change in surface treatments) can disrupt a production process or reduce the efficiency of a system and present very real risks. Third, most materials suppliers cannot overcome "the tyranny of existing infrastructure" (Bridenbaugh 1998). Most industries are based on the design of subsystems and parts, all of which have their own needs for materials and their own supply chain. The complexity of the supply chain makes it difficult to implement a change.

Recommendation 3-2. Materials-supply companies should be encouraged to conduct materials/process R&D. Three potential methods that should be investigated are: mechanisms for larger original equipment manufacturers to assist and encourage materials suppliers to conduct R&D (e.g., guarantees to use the new technology); government programs, such as the Advanced Technology Program, to help defray some of the costs of industrial R&D; and tax incentives to encourage investments in R&D and reduce the risk to the supplier companies.

Parts Suppliers

The linkages between OEMs and part suppliers are generally considered to be the strongest in the materials-production cycle. Parts suppliers are usually purveyors of particular manufacturing technologies that convert the semifinished materials produced by the materials suppliers into finished components ready for installation into final products. Parts suppliers are predominantly contracted by OEMs to make specific parts and subassemblies according to approved specifications and procedures. For example, Howmet, the jet-engine parts supplier, buys components of a superalloy material from materials suppliers and casts the material into single-crystal turbine blades for GE Aircraft Engines and Pratt and Whitney for insertion in their engines.

For many advanced technologies, linkages between OEMs and parts suppliers are predominantly technological oligopolies, with a steady-state number of suppliers for most mature industries of approximately three. Although no deliberate attempts are made to limit the number of suppliers, OEMs tend to have difficulty supporting and managing more than three; fewer than three leaves OEMs at too great a risk of supplier shutdowns or disruptions. In the jet-engine supply chain, for example, there are typically no more than three superalloy producers, titanium producers, forgers, and foundries servicing the industry. The suppliers are almost entirely dependent on the OEMs for their survival and are responsible for producing a significant fraction of the technological content and the majority of the weight of the OEM's product.

Although parts suppliers would seem to enjoy certain privileges and opportunities to profit from this arrangement, there is little evidence that they have benefited. Instead, the suppliers to the jet-engine producers, for example, seem to exist in an unhappy state of "life support," desperate to diversify in "good times" and fiercely competitive in down times. One reason for this is that OEMs are being increasingly pressured by product end-users who demand greater value in a competitive marketplace. This pressure is felt throughout the supply chain.

Because parts must meet precise specifications defined by the OEMs, the strongest links in the relationship tend to be between the design and engineering elements of the OEMs and the corresponding elements of parts-supplier organizations. In the electronics industry, for example, an enormous amount of information is exchanged between the magnetic-head or chip-manufacturing industries and their parts suppliers to ensure that the suppliers' products meet the needs of the OEMs. The high level of standardization of many features (e.g., inputs, outputs, and performance indicators) strengthens this relationship.

Although linkages between OEMs and parts suppliers are strong, the conflicting needs for new, yet totally reliable, technologies can strain the relationship. OEMs generally do not consider themselves developers of supplier infrastructures for new materials/processes. In fact, because of economic concerns and potential liabilities, most OEMs have instituted rigid purchasing systems with known and approved parts suppliers and are skeptical of technologies and suppliers that do not have track records of supplying high-quality parts in high volume.

A discontinuous (i.e., revolutionary) technological change is more problematic than a continuous (i.e., evolutionary) technological change because incumbent suppliers often cannot incorporate the new technologies and produce the new components. Although the configuration of the overall supply network does not substantially change, a discontinuous change in technology often means that incumbent suppliers must be replaced with new, equally reliable suppliers. OEMs often delay incorporating a new technology until the technology and supplier infrastructure has been developed for other products. For example, the use of engineered plastic components for interior/exterior trim on passenger cars and trucks was delayed while the supply industry gained experience with other industries.

Mature industries (e.g., the jet-engine and automotive industries) also have greater difficulty incorporating new technologies than developing industries (e.g., the computer industries). The opportunities for implementing substantial changes in developing industries are numerous as the technology matures and efficiency increases. Once industries become more established and materials/process technologies have been optimized, however, OEMs tend to become assemblers and to reduce R&D on new technologies in favor of evolutionary process improvements. Note the similarities, for instance, between the first 30 years of progress in the automotive industry, when great leaps in technology were made and new records for production and vehicle speed were constantly being set, and the computer industry over the past 30 years. As the automotive industry matured, however, increases in speed and efficiency have become much more difficult to attain.

OEMs urge subassembly and parts suppliers to conduct R&D in technologies for incremental improvement in processes to improve the performance of their products and reduce their costs. On the one hand, parts suppliers are often reluctant to conduct joint R&D projects with OEMs because of the problems involved in convincing OEMs to incorporate new techniques into their products. On the other hand, suppliers are also reluctant to conduct R&D on their own. First, industry's demand for supplier-base reliability can best be met by a small, but not single-source, supplier base. Thus, any innovation a supplier discovers might have to be shared with competitors to ensure that sufficient sources are available to OEMs. Second, OEMs are usually under no obligation to adopt a new technology once it has been developed, thus increasing the risk to the parts supplier. Third, supplier industries usually have large capital investments in processing technology, which increases the costs of introducing new technologies into the market and retards innovation. Because of the high cost of capital equipment, the implementation of new processes and materials can only be accomplished if they can be used on the existing manufacturing tool set. If higher levels of systems integration are required and product liability is increased, this technology lock-in becomes even more entrenched. For example, the consumer electronics industry has fewer problems with technology lock-in than the jet-engine industry because of the higher modularity of computer systems and the lower liability in the event of failure. Finally, if the time required to test and certify a new material/process approaches the limits of the patent-protection period, a company may not have time to recoup its R&D investments before its competitors can legally exploit the technology. Thus, the parts-supply industry tends to be biased toward technologies that are more developed and can be implemented quickly.

Recommendation 3-3. Parts-supply companies should be encouraged to conduct materials/process R&D. Three potential methods that should be investigated are: mechanisms for larger original equipment manufacturers to assist and encourage parts suppliers to conduct R&D (e.g., guarantees to use the new technology); government programs to help defray some of the costs of industrial R&D; and tax incentives to encourage investments in R&D and reduce the risk to the supplier companies.

Recommendation 3-4. Consideration should be given to extending the period of patent protection, especially for applications that require extended certification periods.

Industrial Research Organizations

Many of the companies in the industrial sectors that were studied in preparation for this report (i.e., jet engines, automobiles, and computer-chip and information-storage computer components) conduct internal R&D to provide competitive advantages for their future products. The committee found that the industries represented at the workshops sponsored very little Phase 0 MS&E research and that most of their funding was directed toward meeting their short-term needs. Although this focus on development rather than research may shorten the time from invention to product implementation and may lead to evolutionary product improvements, it does not provide the innovative impetus for the development of revolutionary products for the future.

This has not always been the case. For example, in the recent past, strong basic MS&E research was conducted at large industrial laboratories, such as AT&T (Bell Laboratories) and IBM. This basic research provided much of the technology and materials for the semiconductor and information-storage industry to grow into economic powerhouses. The current electronics industry is an outgrowth of basic research conducted at Bell Laboratories that led to the invention of the transistor in 1948 and the fabrication of the first integrated circuit at Texas Instruments in 1955. These developments also resulted in the formation of new tooling and materials companies to provide production infrastructure and an increase in academic research. The research conducted at industrial laboratories was necessarily multidisciplinary and provided industry with strong patent portfolios to protect their innovative products. It also provided in-house sources of expertise that could quickly address and solve fundamental problems encountered during implementation and accelerated the introduction of new technologies.

Many industrial participants at the workshops recognized that the downsizing of corporations and refocusing on the short-term horizon of stock markets in the 1980s and 1990s had substantially affected the ability and willingness of industry to fund exploratory research. The trend has been for industry to reduce long-term, in-house R&D and to look to academia to fill the void. Industry has also become more involved in industrial consortia to pool research dollars and share results. Although some of these consortia have a long-term vision, most of them are still focused on short-term goals. Relying on university research and consortia also has some drawbacks: the coordination of collaborative projects, the communication of results, and the negotiation of intellectual property rights can be time consuming, problematic, and contentious.

Recommendation 3-5. Industries should establish funding mechanisms and improve its methods of communication and collaboration to support precompetitive, long-term, high-risk research at industrial laboratories, with the participation of academic researchers and suppliers.

Recycling and Disposal

Linkages between the OEMs and the firms that refurbish or recycle products, assemblies, subassemblies, components, and materials are becoming increasingly important—both economically and technologically—as so-called "take-back" regulations spread from Europe to the United States. Take-back regulations require that manufacturers take back their products after consumers are through with them and refurbish and reuse the components or recycle the materials. These regulations will increase the flow of used materials back into the economy and will raise a number of new challenges, such as designing materials so that they can be easily reused. For example, the inclusion of heavy-metal stabilizers and polybrominated fire-retardants in the molding resins used in current computer casings inhibits the recycling of the material when the product is returned.

Because of the scale and complexity of current economic and technological systems, MS&E and end-user communities will have to be more aware of, and concerned about, life-cycle patterns of material use beyond simple disposal and recycling. Material technologies that are useful and benign at a small scale or in the context of a laboratory pilot process can have social, economic, and environmental implications in practice that must be taken into account by materials professionals. Regulatory initiatives focused on specific materials or applications can disrupt product and process designs that would otherwise be economically and technologically feasible, resulting in potentially substantial economic penalties.

The scope and potential impact of regulatory initiatives varies widely. For example, several European countries are considering bans on polybrominated fire-retardants in plastics, which is an important but specific material application. At the same time, they are being urged by environmental groups to ban the commodity plastic PVC altogether. Materials professionals, particularly those working for or collaborating with industrial interests, must be aware of these types of potential changes for new materials and designs to be economically viable. More important, the MS&E community should bring its expertise to bear on social and legal decisions involving materials choices and technologies.

In the highly evolved, complex, service-dominated economies characteristic of developed countries today, it is becoming increasingly important for materials professionals to be sensitive to the social, economic, and environmental context within which materials and products are designed, produced, used, and managed at the end of their life cycles. Fortunately, the developing field of industrial ecology is based on a life-cycle, systems-based view of materials from initial acquisition; to formulation, processing, and manufacturing; to distribution as a material or part of a product; to operational use; to recycling as part of a refurbished product, assembly, subassembly, component, or material; and, eventually, to disposal as waste. The failure to consider all stages of the material life cycle can result in a technology that may be desirable, technically suitable, or manageable at a small scale or in certain uses but that may have substantial social costs at a larger scale or in actual commercial use. Two illustrations are the use of arsenic and silver in the United States (see Box 3-1).

The MS&E community can also make significant contributions to the rational selection and use of materials in the recycling stage of the life cycle. First, the MS&E community can help end-users and the public understand when recycling is, in fact, a good idea, and how optimal networks can be designed. For example, it would be environmentally wasteful (in terms of transportation energy consumption and emissions) if the use of refillable glass bottles results in empty glass containers, which are quite heavy on a volumetric basis, being shipped long distances for refilling. Similarly, shipping lightweight plastic containers long distances for materials recovery to meet a recycling requirement would also be wasteful because significant transport resources would be used for minimal material recovery. Therefore, although materials recycling may be a good idea in general, specific circumstances of recycling determine whether or not it is advisable.

Optimal recycling also requires knowledge of available technologies, for which the expertise of the MS&E community is invaluable. In general, many recycling technologies are fairly primitive, reflecting the fact that virtually all R&D has been directed toward the front end (e.g., material processing, selection, and use) rather than the end-of-life materials management. Thus, for example, the material content of a personal computer—from the circuit board and chips to the solders to the plastic and metal components of the case and ancillary assemblies—has been carefully selected and designed. The end-of-life fate of a personal computer, however, is usually simple disposal in a landfill or, at best, shredding of the product in a hammer mill, followed by secondary smelting of the materials stream to recover metals. As the high social costs of this primitive treatment of materials and products at the end of life, ranging from the waste of potential material streams to toxic effects on humans and ecosystems, are realized, the incentives for the development of more efficient end-of-life material management technologies will grow. The MS&E community will be a critical contributor to the development of these technologies.

Knowledge of industrial ecology is no longer a luxury but a necessary component of technology development that must pervade all of the linkages in the value chain for the materials-production cycle (see Figure 3-2). Industrial ecology is not yet widely taught as part of the traditional MS&E curriculum, however. This deficiency is partly a reflection of the time lag between the rapidly changing social and industrial climate and the traditional MS&E academic focus on the purely scientific and technological dimensions of materials. Industrial ecology is still a young field, and industrial and academic MS&E professionals could make valuable contributions to its development.

Recommendation 3-6. To ensure the appropriate design, production, use, and end-of-life management of materials and products in the future, industrial ecology should be made an integral part of the education and expertise of both MS&E researchers and product designers.

INDUSTRY-UNIVERSITY LINKAGES

The committee found that the linkages and interactions between industries and universities were critical. Barriers to effective interaction range from differences in ultimate objectives to product cycle times. In this section, the committee describes the differences in the fundamental principles of industry and universities.

The role of universities in industrial research has become increasingly important. Universities conduct a broad spectrum of R&D throughout Phase 0, Phase 1, and Phase 2 of the materials/process development timeline and even assist in Phase 3 development as subcontractors or entrepreneurs (e.g., research parks, campus-based industrial-segment research centers, start-up companies, consultants). For example, university researchers have been instrumental in developing process-modeling systems to optimize materials production (Olson, 1998).

The committee found that the relationships between industry and universities are in the midst of a fundamental readjustment. First, industries have been reducing their long-term, in-house basic and applied research in favor of short-term development. As a result, industry has increasingly looked to universities as a source of long-term research. Second, universities are apparently increasing their

---

BOX 3-1
Arsenic and Silver-Laced Water

Every material has a life cycle. Ingredients are formulated, processed, and manufactured into high-tech and low-tech materials or directly into products. These are distributed, sold, and otherwise used until they can no longer serve their original purposes. They are then refurbished, recycled, or used for some other purpose. Sooner or later, the materials end up as refuse to be discarded or managed as waste.

Materials scientists used to be concerned almost exclusively with the early phases of a material's life cycle. Keeping costs down while maintaining marketable quality were the major goals. But the latter phases of the materials life cycle have been slowly infiltrating the general mind-set of the materials community. Creating new, affordable, more capable materials is no longer enough. New drivers to minimize negative social, economic, and environmental consequences of materials throughout their life cycles have become part of the equation. The following examples suggest the new kinds of cognitive skills necessary to adapt to the cradle-to-grave perspective.

For the past 30 years, the United States has used about 20,000 metric tons of arsenic annually—about two-thirds of the world's arsenic consumption. In the past, the major uses of arsenic, including pesticides and drying agents, were dispersive, and the arsenic was essentially unrecoverable. Now, the toxic metal is heavily regulated, and its use in obviously dispersive applications has been considerably curtailed.

Still, arsenic-bearing compounds have been widely dispersed into the environment through an unexpected channel. Each year, 5 billion board feet of pressure-treated wood are protected from termite damage and dry rot using chromated copper arsenate, which accounts for 90 percent of worldwide arsenic demand. These agents have almost completely replaced organic wood preservatives like creosote.

On small scales, arsenic compounds would not be troublesome. But arsenic-based preservatives have become the lumber industry's standard. Every year, 15 cubic miles of arsenic-containing materials diffuse across the landscape in the form of architectural framing, decks, and hundreds of other structures. As a result, a toxic metal continues to be dispersed throughout the environment, and there appears to be no simple or inexpensive way to recover it.

---

pursuit of industrial funding, either because of an overall decrease in government funding for MS&E R&D or because of the general reallocation of government R&D funding to other important fields (e.g., biomedical research) or because of a general increase in the number of MS&E researchers applying for grants (which has increased the competition for government funds). Of course, the pursuit of industrial funding by universities could also reflect a genuine desire on the part of university researchers to see the results of their Phase 1 and Phase 2 research implemented. In any case, universities are focusing more on short-term industrial

Silver, another usually prized metal, has also been causing problems at the end of its life cycle. Several years ago, the California Regional Water Quality Control Board discovered high levels of silver, an aquatic toxicant, in sediments and biological systems of the San Francisco Bay. The source of the contamination came as a surprise. About half of the silver flowing through the U.S. economy is used in the form of photographic fixer solutions in dentists' offices and photography laboratories. Researchers found that most of the fixer, along with its silver, was literally going down the drain, and from there some of the silver was finding its way into the bay.

Virtually all of the experts—from the scientists and engineers who created the fixers and accompanying photography-based technologies to environmental scientists and activists to government regulators—had no idea that so much silver was being poured down the drain. In fact, the environmental regulations created to curb the introduction of silver into the environment had contributed to the problem. By imposing a relatively high fixed cost on residual streams containing silver, the regulatory structure inadvertently created strong incentives for not recycling contaminated waste streams. For individual dentists or small photography shops that, individually, use small amounts of fixer, pouring used fixer solution down the drain is a rational business decision that doesn't seem to do too much damage. Indeed, if only a few of them disposed of fixer that way, there would be little environmental problem. Collectively, however, these many small amounts of silver add up to an environmentally damaging influx of silver to the environment.

Materials scientists and engineers, as well as environmental professionals and regulatory communities, have tended to look for environmental solutions in the early phases of materials life cycles—the manufacturing processes and manufactured products. But the example of silver contamination suggests that social, regulatory, cultural, and other factors also influence how products ultimately flow through the environment.

Countless other scenarios just like these are undoubtedly being played out. To identify them and prevent new ones, the materials community must focus on how materials are actually used and handled once they leave the factory. Otherwise the regulatory framework and the human penchant for convenience may lead to practices that undermine even the best efforts to minimize the environmental costs of providing the materials society needs or wants.

research than they did previously, and they are becoming increasingly involved in filing for patents and in other business-oriented activities.

Although the industrial representatives at the three workshops emphasized that universities should continue to conduct some long-term R&D to support the innovation pipeline, they also were of the opinion that the increase in short-term research at universities was advantageous for both parties. It has reduced the overall cost of research for industry, strengthened the linkages with universities, and produced new avenues for introducing innovations. Most important (from

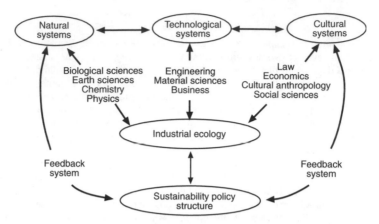

FIGURE 3-2   Schematic representation of the linkages associated with industrial ecology. Source: Allenby 1999.

industry's perspective), however, is that the emphasis on short-term research has somewhat shifted university research toward "real engineering" programs. One remaining concern about academic MS&E R&D was that more government funds were being invested and more graduates were being produced in low-volume, specialty materials and processes than in the high-volume materials and processes on which many traditional companies rely.

University representatives at the workshops expressed their concern that the trend toward short-term research could pose considerable risks to universities. First, focusing on short-term research could limit the resources available for long-term research, which is necessary to maintain the innovation pipeline. Therefore, many of them felt that industry should assume more responsibility for long-term research, which will ultimately determine the competitiveness of U.S. industry. Second, universities are concerned that focusing only on short-term industrial research could undermine their credibility as nonprofit organizations whose purpose is education. Many in the academic community are concerned that the realignment with industry could interfere with the main missions and educational goals of universities. Universities are in a difficult position because, as the field of MS&E broadens, they must train graduates for careers in a wide spectrum of industries.

The committee found that many of the concerns about relationships between industry and academia could be alleviated by better communication and data sharing, more compatible equipment, a stronger policy for interaction, more industry access to research results, more compatible time scales, and more compatible objectives and reward schemes.

Weaknesses in Communications and Data Sharing

The first weakness in industry-university linkages is in the communication of information. Because of proprietary concerns, industry tends not to disclose all of the information about the context of the research problem and the complexity of the final product for which the technology will be used. Therefore, university researchers often find it difficult to define the problem correctly, and, ultimately, industrial partners may be dissatisfied with the results. By improving communications, researchers would become more interested and involved in the problems of high-volume materials and could design experimental programs that produce more useful results for industry and are more appropriate to the university paradigm. The problem arises, in part, because the supply industries, which tend to be confronted with most of the fundamental material/process problems, have historically been less involved in funding university research. Another reason for the problem is that the decrease in Phase 0 and Phase 1 R&D by industry is attracting fewer science graduates to industry, who were the traditional links between industries and universities.

Recommendation 3-7. Industry should improve its communication of information about technical objectives, context, and product complexity to universities to ensure that university research is applicable and properly focused. Increasing adjunct professorships for industrial scientists and engineers at universities and developing arrangements for Ph.D. and masters students to conduct their research in industry facilities would increase and improve personal interactions between the two communities.

Production Scaling

The second weakness in industry-university linkages is the incompatibility between industrial and university equipment, which adversely affects the usefulness of research results. Academic researchers should have access to state-of-the-industry equipment for conducting and verifying their research. The development of applicable process models could alleviate this problem by improving the general understanding of the fundamental principles of new systems and determining industrial parameters for the successful transition of laboratory-developed materials/processes to industrial-scale production. For example, academic researchers could be given access to processing equipment similar to production-scale equipment and process models that represent the variability in industrial environments.

Recommendation 3-8. Universities should work with industry to develop methods to increase the applicability and improve the reliability of laboratory data and to demonstrate the potential for scale-up of new technologies.

Policy for Interaction

The third weakness in industry-university linkages is the lack of a standard policy and procedure for interaction. Universities devote considerable time and resources to establishing links with industry and developing contracts that include intellectual property rights and the licensing of new technologies. However, no standards have been developed defining the responsibilities of all parties and eliminating the need to reinvent contracts and contacts with each new project.

Recommendation 3-9. Standard university-industry contracts for sharing intellectual property rights and licensing new technologies should be developed. These contracts should clearly define the responsibilities of all parties. Standard contracts would reduce the time and legal costs required to establish industry-university research programs. To ensure that standards were acceptable and equitable to all parties, they should be approved by industry, academia, and government (e.g., professional societies, academic deans of research, and high-level government funding organizations).

Industry Access to Research Results

The fourth weakness in industry-university linkages is the inaccessibility of many university research results. As described in Chapter 2, the results of most basic research programs are disseminated via academic conferences and journals. However, the number of experts in industry who can evaluate this information and assess innovations is decreasing as industrial basic research declines. The remaining industry researchers have less time and fewer resources to keep abreast with new developments than they had in the past. Therefore, technological innovation might not attract their attention, and promising innovations may be overlooked.

Recommendation 3-10. Industry should establish methods for identifying and assessing materials/process developments from universities and disseminating the results to industry.

One possible method for improving industry's access to university research is through the development of a nonindustrial, worldwide-web-based research clearinghouse that could make the results of independent research easily searchable and thus more accessible to industry. However, this method would not address industry's problem of the lack of expertise and resources to assess this information. An alternative method might be for consulting companies to assess research results in a given field and bring the consequential innovations to the industry's attention. One advantage of this method would be that linkages could be established before the research results were published when industry could take full advantage of the innovation.

TABLE 3-1 Characteristic Time Scales for Academia and the Automotive Industry

Academia	Automotive Industry
2 years: capital budget cycles	1 quarter: shareholder profit expectations
2 years: Masters of Science project	1 year: budget cycle
3 years: typical government grant (5 years for centers of excellence and NIST's Advanced Technology Program)	1–3 years: typical automotive grant to university (1-year grants renewable at automaker option)
5 years: Ph.D. project	2 years: typical Phase 3 horizon
6 years: tenure probation period	3 years: sign-off to production
Lifetime: disciplinary focus (tenure outcome)	4–10 years: typical production run

Differences in Time Scales

The fifth weakness in industry-university collaboration is the frustration caused by differences in time scales and process cycles (e.g., Table 3-1). The critical and most variable delay is in moving from material concept development (Phase 1) to product integration and sign-off (start of Phase 4, end of Phase 3).[2] As Table 3-1 shows, an industrial concept, such as a new vehicle concept, is developed over the two-year period before sign-off, and during this period the most intensive consideration is given to new technologies (e.g., the use of tailor-welded blanks or hydroformed tube chassis). Although production tooling and procedures must still be developed, all of the technologies used in the vehicle must be ready for implementation at sign-off, with complete economic justifications and selection of suppliers. Even if an attractive new technology appears within a month or two after sign-off, it usually cannot be included in the product. Universities, however, cannot operate on a two-year cycle and still educate students. Unlike industry, whose primary research objective is to develop new

---

[2] For some purposes, such as patent lifetime and cost recovery considerations, it may be useful to include the time from Phase 4 to production (or profit making). From this perspective, the typical product production lifetime and time to sign-off-for-production are both relevant. A production-ready process will remain on the shelf until a new product passes sign-off and the product enters production. This factor is actually less important in the automotive industry than in the aerospace industry, because of the many product lines that must be redesigned and produced, perhaps an average of one per year for each manufacturer. In the jet turbine industry, however, it may be many years between new product sign-offs. In the electronics industry, however, most of the technology is developed in response to a market pull, as embodied in an industry road map, so much of the new material/process comes to Phase 3 quickly.

technologies or improve existing ones, university graduate students are required to teach and be taught as well as to conduct research. Students are also largely unknowns as they enter graduate programs, and most projects rely on the work of a small number of students (so there is little statistical evening). Furthermore, projects must be considered, designed, proposed, and developed before the student arrives on campus so equipment can be designed and materials and supplies ordered. The longest step in the preproject timeline is usually project design, proposal writing, and consideration of the proposal by funders of the project.

Recommendation 3-11. Industry should develop mechanisms to coordinate industry-sponsored research with university research cycle times without compromising university or industrial missions and timelines.

Differences in Objectives and Reward Schemes

The sixth weakness in industry-university linkages is differences in motivation. An academic R&D program requires not only funding and equipment but also a consensus that it fits into the academic culture and is in keeping with the educational mission of the university. A common problem encountered by universities is evaluating junior faculty members engaged in industrial R&D. Tenure appointments are generally based on the publications of the candidate and the evaluations of recognized faculty members at other institutions. Industry-imposed limitations on publishing the results of research in the open literature or on collaborating with other faculty members puts junior faculty members at a disadvantage for tenure.

Recommendation 3-12. Academic administrators should consider the value of industrial (and other nonacademic) interactions typical of industrial research in their faculty evaluations.

Relationships between industry and universities also have an educational component. Industry relies on universities to educate technical and management personnel. Therefore, industry is concerned that the current MS&E curriculum is turning out graduates with narrowly focused knowledge of materials that are currently of little economic consequence instead of graduates with a broad general knowledge of the materials that are the mainstays of industrial competitiveness. Better communication between industries and universities could help determine an appropriate balance between materials innovation and industrial relevance.

Recommendation 3-13. Industry and universities should develop mechanisms to increase personal interactions and communications and to determine an appropriate balance of training and education to ensure the continued success of the MS&E R&D community, as well as satisfying the needs of industry.

Potential mechanisms for increasing personal interactions include (1) increasing adjunct professorships for industrial scientists and engineers; (2) encouraging joint research projects; (3) increasing the flexibility of exchange programs between universities and industry to allow representatives of either community to spend as much time as necessary and appropriate (e.g., from single day visits to full year sabbaticals); (4) organizing seminars and workshops to introduce university faculty members to the complexities, intricacies, and economics of manufacturing; and (5) enabling students to conduct research in industry (e.g., cooperative programs that provide both undergraduate and graduate students with opportunities to work in industry prior to graduation).

Mechanisms to Improve University-Industry Interactions

Industry and universities are reexamining their relationships. University programs that have revised their research agendas based on the problems identified by their industrial partners are finding it easier to find industrial partners, secure funding, and, presumably, facilitate the adoption of research results. This new market-driven research agenda is in stark contrast to the more traditional, independent, idea-driven research of single-investigator, university laboratories. In the traditional climate, which works extremely well for developing basic knowledge and preparing students for careers in basic or academic research, students conduct curiosity-driven research in relative isolation, using university laboratory space and equipment, and with minimal concerns about the practical application of their work.

The center of excellence is a new model for university research that is rapidly gaining acceptance. Centers of excellence, in sharp contrast to the traditional model of university research, have a clear research focus, involve collaboration by several faculty members (often from different disciplines), provide shared facilities, and have proactive industrial outreach programs. Interdisciplinary teams are better able to meet the needs of industry for relevant university research. The advantages of a center of excellence over the traditional model include: (1) it creates a critical mass for the rapid exchange of information; (2) it identifies industry segments interested in specific research projects; and (3) it provides investigators with greater access to the increasingly expensive and sophisticated equipment required for materials research. A center of excellence provides industry with a single location from which to anticipate relevant research results and a pool of recruitable students with immediately applicable skills and experience working in teams. Centers are also better able to respond to multidisciplinary federal research initiatives that require industrial outreach (e.g., the National Science Foundation's Materials Research Science and Engineering Centers and Science and Technology Centers Program).

Centers of excellence commonly recruit industrial participants using an established fee structure and a common intellectual property agreement.

Membership in a center provides industry with access to the output of all of the research performed at the center, which may have a research budget 10 to 100 times the membership fee. Research results can be shared with industrial members through activities such as on-campus research reviews and workshops, faculty visits to member sites, and student internships in industry. Participation in research programs supported by industrial consortia can provide a venue for university/industry collaborations and facilitate efforts by new faculty to establish research programs by providing them with access to well equipped facilities.

Recommendation 3-14. Universities should consider establishing centers of excellence as a mechanism for "marketing" their research, promoting customer-oriented research at their universities, improving the chances of successful technology transfer, and improving linkages to industry.

INDUSTRY-GOVERNMENT LABORATORY LINKAGES

Government laboratories also play an important role in industrial research because they conduct a broad spectrum of R&D throughout Phases 0, 1, 2, and 3 of the materials/process development timeline. The committee found that the relationship between industry and government laboratories has changed substantially in recent years.

Changes in government policy since the end of the Cold War have resulted in significant changes in government laboratories. For example, U.S. Department of Defense (DOD) laboratories previously conducted a great deal of MS&E research related to the development of new weapons platforms and equipment (e.g., new stealth fighter planes). Since the end of the Cold War, however, DOD has been more concerned with maintaining current capabilities then developing new ones and now relies on industry to lead materials production and R&D.

The same is true of the U.S. Department of Energy's (DOE) national laboratories. At the end of the Cold War, the three large DOE defense laboratories (Los Alamos, Lawrence Livermore, and Sandia) were directed by the government to refocus their research programs on industry needs. The national laboratories faced many of the same barriers to working with industry as universities (e.g., different motivation, intellectual property rights issues, and cumbersome contracting procedures). Nevertheless, over a period of five or six years, many cooperative R&D projects were initiated. At the same time, the seven multiprogram civilian DOE laboratories increased their industrial cooperation. In the mid-1990s, the three DOE defense laboratories redefined their defense missions, focusing on the stewardship of nuclear weapons and nuclear nonproliferation. At about the same time, the federal government decreased its support for cooperative work with industry.

Most of the research previously conducted at the DOE weapons laboratories was not directly relevant to industry. Industrial representatives at the three workshops suggested that an increase in short-term research at federal laboratories would be beneficial for both industry and the laboratories (for much the same reasons as for university laboratories). However, most also believe that the laboratories should continue to conduct some long-term R&D to maintain the innovation pipeline. In addition, they recommended that the peer-review process for DOE laboratories be augmented to ensure the quality of the research and the applicability of results to the needs of industry.

Like their university counterparts, laboratory representatives expressed their concern that the general trend toward short-term research and greater alignment with industry would move the laboratories away from their main mission of long-term research.

Recommendation 3-15. The federal government should continue to encourage interaction and communication between federal laboratories and industry and to establish partnerships, in keeping with laboratory missions, in areas that will benefit industry.

Potential mechanisms for increasing personal interactions include fostering more joint research projects; increasing the flexibility of exchange programs between government laboratories and industry; and organizing seminars and workshops to introduce government laboratory personnel to the complexities, intricacies, and economics of commercial manufacturing.

Most industry representatives at the jet-engine workshop were extremely concerned about changes in the DOD laboratories. For example, the domestic jet-engine industry has been closely linked with, even reliant on, basic materials/process R&D conducted and funded by the Air Force. Most major improvements in the efficiency of jet engines have resulted from DOD initiatives funded and/or conducted by the Air Force, which also provided the basis for implementation, reliability testing, and scale-up. In a dramatic reversal of roles, the Air Force now relies on industry to lead materials/process research initiatives. The industry, however, which has just emerged from an extended period of low profitability, severe downsizing, and reorganization, cannot support these initiatives. Industry representatives feared that, without the support of the Air Force, no long-term research would be conducted and that the competitiveness of the domestic jet-engine industry would suffer. In short, the jet-engine industry believes it is in the national interest for DOD to continue to support basic materials/process research and to remain closely linked with the domestic industry, while DOD representatives believe that industry should assume greater responsibility for long-term research because it would be in its own best interest.

INDUSTRY-GOVERNMENT LINKAGES

The relationship between government and industry is extraordinarily complex, but there are three main methods by which government affects industry: direct funding of R&D; business regulation; and environmental regulation. Government regulation of business (e.g., liability, international trade, antitrust, and tax legislation) is beyond the scope of the committee's charge and expertise and, therefore, is not discussed further. This section focuses on the effects of environmental regulation on industrial materials and process development.

Environmental regulations can compel industry (1) to modify or replace an existing manufacturing process or production facility to reduce harmful emissions or (2) to modify or augment a product design to improve safety or reduce harmful emissions. Either of these changes can cause manufacturing delays and add to the cost of materials implementation. More important, however, replacement technologies must not only satisfy government regulations but must also maintain required quality and performance levels. Regulatory changes also affect government operations. For example, continued changes in standards and regulations can cause backups in permit approvals, which can slow the implementation of new technologies.

Although, in general, industry is opposed to government interference in commerce, the committee found that industrial participants in the workshops did not believe that product regulation was a major deterrent to industrial competitiveness because all companies must comply equally with new regulations. In fact, regulation can stimulate innovation by motivating companies to conduct cooperative, precompetitive research and by helping them overcome the cost barriers that limit the introduction of new materials/processes. Government regulation can also limit liability in certain industries. In the aerospace industry, for example, industry and the Federal Aviation Administration (FAA) tend to see their relationship as a partnership with respect to the introduction of new materials and processes. By working closely with the FAA, the aerospace industry can ensure that safety issues and liability concerns are fully addressed.

Recommendation 3-16. Government regulatory agencies and the industries they regulate should attempt to change the current regulatory climate to mutually constructive cooperation and goal setting to promote the adoption of new materials that further societal goals.

Many government agencies fund Phase 0 and Phase 1 materials/process R&D. For example, the National Science Foundation funds basic research and education in science and engineering, principally in academia. DOE and DOD have similar programs to fund Phase 0 and Phase 1 R&D. In the past decade, as federal programs have focused more on the development of precompetitive technologies (e.g., improving automotive fuel economy and reducing pollution), more funding has been used for Phase 2 R&D. Many state governments have also established programs to support technology areas as a way of attracting new high-technology businesses to their states.

The *Technology Reinvestment Program* of the Defense Advanced Research Projects Agency (DARPA) was a four-year program to shift DARPA's defense-oriented manufacturing research to a more commercial-industry-oriented program. The program was managed jointly by DARPA, the National Science Foundation (NSF), and DOE. The program recognized that DARPA would be less able to support and implement cutting-edge manufacturing technology research as defense budgets decreased.

The *Advanced Technology Program* (ATP), sponsored by the National Institute of Standards and Technology, is intended to benefit the U.S. economy by stimulating the development of innovative technologies at the preproduct stage. Joint ventures must account for at least half of the project costs, and single companies are required to pay all of their indirect costs. Universities may participate in joint ventures or as subcontractors. Funding for ATP has averaged around $200 million per year for the past five years. ATP has established 17 focused programs, seven of which are principally oriented toward materials or processing.

DOD supports manufacturing technology through the *Manufacturing Technology Program* (ManTech), which supports 15 centers of excellence in manufacturing fields ranging from apparel to electro-optics. ManTech also funds the Best Manufacturing Practices Center of Excellence to make the results of R&D at the centers and other defense-related industry knowledge available to industry at large.

The *Partnership for a New Generation of Vehicles* (PNGV) is a partnership of 20 federal laboratories and Chrysler, Ford, and General Motors to improve U.S. competitiveness in automotive manufacturing through the evolution of an environmentally friendly car with triple the fuel economy of today's midsize car. Seven agencies and the automakers jointly fund PNGV, and DOE directs the program. Materials and manufacturing are main areas of investigation.

Four programs of the NSF Directorate of Engineering are noteworthy for their interaction with industry. *Industry/university cooperative research centers* (I/UCRCs) leverage a modest investment by NSF into a focused cooperative research program with industry support. More than 25 I/UCRCs have been established in the past 15 years. They represent one of the best examples of industry-university interaction and cooperation. *State/industry university cooperative research centers* (S/IUCRCs) extend the I/UCRC model, focusing on state or regional economic development, often including proprietary projects with both industry and state support. *Engineering research centers* (ERCs) represent an integrated university-industry focus on complex engineered systems. Two existing ERCs, at Purdue and Ohio State, focus on manufacturing. The *Grant Opportunities for Academic Liaison with Industry* (GOALI) program brings individual engineering faculty members and industry into close working contact. The GOALI program provides funding for industry engineers to work in academia on collaborative projects.

The *Industries of the Future* (IOF) Program was established to help the DOE Office of Industrial Technology leverage government and private funding by focusing research on industry-developed visions and technology road maps (NRC, 1999b). The objective of the IOF program is to improve government-industry partnerships, ensure the relevance of research projects, encourage industry participation, and facilitate the commercialization of new technologies. The long-term goals are a 25-percent improvement in energy efficiency and a 30-percent reduction in emissions for the IOF industries by 2010 and a 35-percent improvement in energy efficiency and 50-percent reduction in emissions by 2020 (OIT, 1997).

The New York State Science and Technology Foundation is a public corporation that administers a range of financial- and technical-assistance programs designed to stimulate economic growth and job creation in New York through the transfer of technology from the laboratory to commercial application. One of its three main endeavors is the *Centers for Advanced Technology* program, which encourages new and high-technology product and service development through R&D, technology transfer between universities and industry, and education and training.

All of these Phase 2 government/industry/university cooperative programs require significant industry matching funds. Their overall focus is primarily based on meeting industry goals and objectives.

Recommendation 3-17. Federal and local governments should expand their programs to fund joint industry-university research programs to enable new technologies to make the transition from the laboratory to industry. These programs should focus on involving both original equipment manufacturers and suppliers in the selection and management of research projects.

CONSORTIA

The formation of consortia to conduct precompetitive research is a relatively recent phenomenon that started in 1984 with passage of the National Cooperative Research Act. The original objective was to provide a mechanism to enable product manufacturers to coordinate their Phase 0 and Phase 1 precompetitive research in response to foreign competition without violating antitrust laws. Since then, the missions of most consortia have been expanded to include: (1) conducting joint Phase 0 and Phase 1 research on high-risk, precompetitive technologies; (2) obtaining government funding; (3) developing technology road maps; (4) maximizing the value of university research; and (5) acting as industry spokesgroups.

Consortia, which can include major suppliers and manufacturers, applicable university programs, and relevant government laboratories and agencies, are funded by contributions from major participants. Consortia generally have four types of members:

- full industrial members, who pay dues in the tens or hundreds of thousands of dollars and generally have full and immediate access to R&D results, as well as full participation in decision-making processes

- partial industrial members, who pay dues in the tens of thousands of dollars or less and generally have limited access to the R&D results
- research members, who conduct the R&D
- governmental agencies, which provide a large share of the funds for R&D

For rapidly changing, high-profit industries, research may be funded exclusively by industry. The electronics industry, for example, has established several industry-funded consortia to develop visions of the near future and fund R&D projects.

Consortia accomplish their objectives in two ways. First, they provide neutral territory on which competing industries can meet to identify, develop, and maintain the research initiatives most important to their competitiveness. Second, they serve as links among industries and research institutions to ensure that short-term and long-term research initiatives are effective and efficient. The main mechanism by which consortia operate is through industry road maps, frameworks for setting priorities in materials research. Road maps have been very useful for establishing goals and priorities that have led to the development of advanced technologies in newer industries, such as electronics. Some advantages of road mapping are listed below:

- Road maps are high-level mechanisms for identifying and disseminating information about the problems, challenges, and opportunities in a given field.
- Road maps help define the issues facing industries and identify gaps in technology.
- Road maps are communications tools that enable all segments of an industry (e.g., researchers, suppliers, systems integrators, and recyclers) to contribute to the industry's development.
- Road maps bring all segments of the industry into the development process—from fundamental R&D to final assembly—in a coordinated way. Road maps must be sufficiently detailed so that each segment understands the R&D areas to be pursued.
- Road maps based on the input of industries, suppliers, academia, and government represent a consensus on R&D goals and directions. They also provide a way of leveling the playing field among researchers and industries, lowering the overall risk, and ensuring that a market will exist for innovation.
- Road maps are tools for helping funding agencies determine which projects to fund.

The process of developing industry road maps encourages the participation and interaction of experts across institutions and disciplines, which fosters understanding and communication between materials experts and product designers, both within and across industries and research institutions, and minimizes "missed opportunities."

Recommendation 3-18. The MS&E communities should promote the use of road maps (1) to identify the issues facing industries and the gaps in the technology; (2) to serve as a means of communication for all segments of an industry to contribute to the industry's development; (3) to serve as an organizational mechanism to coordinate all segments of an industry; (4) to provide integrative structures through which all segments can "buy into" the goals and research directions of the industry; and (4) to provide funding agencies with the information necessary to manage their research budgets.

The development and implementation of road maps are not free of risk, however. First, an industry that simply follows the schedule stipulated in a road map will not survive. To control or increase its market share and maintain its competitiveness, a company must attempt to preempt its competitors by introducing new technologies before the dates established on the road map. Because of the constant pressure to beat the deadline, road maps are usually obsolete within two years. Thus, road maps must be treated as living documents rather than set guidelines. Unless consortia vigilantly maintain and update their road maps, the competitive advantage they provide will be lost.

Second, road maps could lead to technology lock-in. By necessity, road maps are mainly concerned with evolutionary R&D and cannot identify or support revolutionary innovations. Industry must be careful not to eliminate revolutionary research in the name of efficiency and leave themselves vulnerable to competitors developing leapfrog technologies. Once the industry recognizes the limitations of a road map, however, revolutionary ideas can be developed by veering off the incremental course set by the road map and envisioning leapfrog technologies based on completely different paradigms. The most effective way to avoid technology lock-in is to use road maps to forecast and prioritize needs, not solutions.

Third, road maps can only be truly successful if the participants remain involved and provide conduits for the transfer of results. Road maps must also clearly define precompetitive and proprietary interests to ensure that companies have a basis on which to compete once the R&D stipulated in the road maps has been completed.

4

Priorities

"Forward, the Light Brigade!"
Was there a man dismayed?
Not though the soldiers knew
Someone had blundered:
Theirs not to make reply,
Theirs not to reason why,
Theirs but to do and die:
Into the Valley of Death
Rode the six hundred.

From "The Charge of the Light Brigade"
Alfred Lord Tennyson (1809–1892)

THE PURPOSE OF THIS CHAPTER is to highlight the committee's most important findings and recommendations. The fundamental focus of this report is the importance of materials advances in the development of marketable products. The committee found that, although in some cases the introduction of a new material has revolutionized an industry (e.g., silicon chips for the electronics industry, optical fibers for the telecommunications industry, and titanium for the aerospace and aircraft industries), the vast majority of materials advances have been evolutionary. In either case, it has taken from 10 to 20 years for typical material advances to be widely used. As a result of these long development times, patent protections often expire before the material/process innovators realize significant revenues or even recoup their original investments. This has discouraged the development of innovative materials.

An idealized commercialization process and the many linkages necessary for materials and processing advances to make the transition from the laboratory to the marketplace were discussed in Chapters 2 and 3 of this report. Chapter 2 introduced a conceptual schema for the analysis of the materials development and commercialization process, which includes the following notional phases:

- Phase 0. knowledge-base research
- Phase 1. material concept development
- Phase 2. material/process development
- Phase 3. transition to production
- Phase 4. product integration

Phase 2 is typically the most difficult phase of the development cycle to navigate successfully. The primary objective of Phase 2 R&D is to scale-up production from the laboratory to the prototype level so that the business risks involved in the application of a material/process innovation can be quantified. Phase 2 ends when the innovation has been shown to have both the potential for being scaled up to production level and economic advantages and when an industrial enterprise decides to investigate integrating the technology into a product. In many cases, Phase 2 represents the transition from "technology push" (i.e., research priorities are established by the MS&E community based on technological attractiveness and perceived applicability) to "product pull" (i.e., industry needs and priorities are the primary criteria for further development). Some have described Phase 2 as "the valley of death."

Recent experience has left the user community as well as the R&D community frustrated. Many in the user community are of the opinion that materials R&D has been misguided and preoccupied with exotic but impractical technologies. Many in the MS&E community feel that the fruits of their research have not been adopted and that the user community is overly conservative. At the root of these feelings are technologies that have not made the transition from technology push to product pull.

The importance of Phase 2 R&D and the substantial differences between Phase 2 and traditional Phase 0 and Phase 1 research are gaining recognition with funding agencies, universities, government laboratories, and industry. Overcoming the barriers to Phase 2 R&D is the most promising way to shorten the time to market of laboratory innovations.

The committee identified the following principal barriers to the smooth passage through Phase 2 R&D:

- high development costs
- high technical and business risks
- inadequate communications and education

Any successful innovation must be a cost-effective solution to a real problem. Therefore, the MS&E community must have a good understanding and appreciation of costs in the materials selection process. Focus on technological innovation without regard to cost is unlikely to lead to success. Historically, funding for Phase 2 research has been inconsistent. Although the highest costs of new materials have been associated with process definition and testing, funding often stops before these stages have been reached. The funding gap may result from uncertainty among the MS&E community, industry, federal and state funding agencies, and entrepreneurs over who is responsible for the identification and funding of Phase 2 R&D programs.

Because of the high cost of testing new materials, many materials advances have never been exploited. The use of modeling and simulations to provide preliminary assessments of materials and component performance could help alleviate this problem. However, extensive databases and knowledge-base systems will be essential to effective modeling and simulation.

Economic pressures compel manufacturing enterprises to evaluate the technical and business risks associated with every new technology. As a consequence, mature industries are more likely to fund incremental R&D, for which the risks are better understood. Revolutionary changes are most likely to come from entrepreneurs who are willing to accept higher risks in search of high returns. Different industries perceive risks differently. In the electronics industry, for example, risks are predominately associated with commercial considerations. In the automotive industry, risk may be associated with product safety or the potential for huge recalls. In the jet-engine industry, safety is paramount, and no company will introduce a new material or process unless it has been proven to have a positive or, at worst, a neutral effect on safety. The cost of running long-term, expensive tests to verify product reliability is a major barrier to innovation. Risk assessments and evaluations of all performance criteria can cost tens of millions of dollars, and these costs are major impediments to the introduction of new materials.

In the opinion of the committee, universities are producing MS&E graduates who are technically well educated, but whose focuses are too narrow for the current business climate. Educators should ensure that MS&E researchers and graduates can communicate effectively with producers and designers so that their ideas can be successfully brought to market. Researchers and engineers must understand that producers are looking for simple, robust processes, continuity of demand, and the potential for profit; designers think in terms of life-cycle cost, risk management, and consistent and reliable suppliers.

One way to improve the preparation of MS&E researchers and graduates is to involve research universities, in partnership with industrial researchers, in Phase 2 R&D. However, this has been difficult for the following reasons:

- the multidisciplinary nature of Phase 2 R&D and the wide spectrum of expertise required to complete material/process developments
- the lack of access to industrial-scale equipment
- the evaluation of academic researchers based on refereed publications and invention disclosures
- the incompatibility between industry funding and planning cycles and the time frames required for graduate students

Recommendation 4-1. The MS&E and user communities should focus their efforts on strengthening linkages to bridge the Phase 2 "valley of death" of technology development.

Although there are major differences between industries, some general approaches can be taken to improve Phase 2 R&D. The key to bridging the valley of death is to establish an environment in which innovations are desired and anticipated by those who will use them and business considerations are addressed early in the development process by the MS&E researchers. Focusing on the following areas will improve the chances that materials and processing innovations will be successfully commercialized:

- improving Phase 1 linkages (setting the stage for product pull)
- establishing the potential of an industry for Phase 3 and Phase 4 R&D (getting down to business)

SETTING THE STAGE FOR "PRODUCT PULL"

Even though some innovations have succeeded without a clearly defined need, the committee found that commercialization is much more likely to succeed if product needs drive the innovation. Phase 1 researchers must become more aware of user needs and consider them in designing their research programs, thus establishing a "product pull" (i.e., setting research priorities based on product needs).

Consortia

Many industrial research laboratories have decreased their support for Phase 0 and Phase 1 MS&E research, directing more of their activities toward meeting short-term needs. Although this change in focus could shorten the time for product implementation and lead to evolutionary product improvements, it provides no incentive for revolutionary innovations. To compensate for this lack of incentive, industry has turned to academic researchers and consortia to pool research resources and share results. Consortia, with or without government participation, provide a mechanism for sharing the risks and costs of developing new processes and materials. Consortia provide neutral ground where competing industries can meet to identify, develop, and maintain the research initiatives most important to their competitiveness. Consortia can also serve as links among industries and research institutions to ensure that short-term and long-term research initiatives are effective and efficient.

Industry Road Maps

Industry road maps are the primary mechanisms for establishing research goals and priorities for materials research early in the development process. Road maps have been very effective for the development of advanced technologies in newer industries, such as electronics, and are especially important for the development of complex products. The road map development process facilitates linkages between experts across institutional and disciplinary boundaries. Road maps are valuable for the MS&E community because they can (1) identify issues facing industries and gaps in technology; (2) be used as communication tools to allow all segments of an industry to contribute to the industry's development; (3) act as organizational mechanisms for bringing all segments of an industry into the development process; (4) serve as integrative structures through which all segments of an industry can reach consensus on goals and research directions; and (5) provide funding agencies with the information necessary to manage their R&D budgets.

Centers of Excellence

· The center of excellence is a new model for university research that is rapidly gaining acceptance. Centers of excellence, in sharp contrast with the more traditional model of university research, have a clear research focus, involve collaboration by several faculty members often from different disciplines, provide shared facilities, and have proactive industrial outreach programs. An effective center of excellence (1) creates a critical mass for the rapid exchange of information; (2) identifies industry segments interested in specific research projects; and (3) provides investigators with greater access to the increasingly expensive and sophisticated equipment required for materials research. A center of excellence provides industry with a single location from which to anticipate relevant research results and a pool of recruitable students with immediately applicable skills and experience working in teams. Centers are also better able to respond to multidisciplinary federal research initiatives that require industrial outreach.

Recommendation 4-2. The following three primary mechanisms should be given priority to establish product pull in the early stages of technology development (during Phase 1 and, perhaps, as early as Phase 0):

- consortia and funding mechanisms to support "precompetitive" research (Recommendation 3-5)
- industry road maps to set priorities for materials research (Recommendation 3-18)
- university centers of excellence to coordinate multidisciplinary research and facilitate industry-university interactions (Recommendation 3-14)

GETTING DOWN TO BUSINESS

The successful commercialization of materials and process advances is generally driven by one of four end-user forces: (1) cost reduction; (2) cost-effective improvement in quality or performance; (3) societal concerns, manifested either through government regulation or self-imposed changes to avoid government regulation; or (4) crises. Without at least one of these drivers, industries that use materials have little motivation to implement technological advances. However, the importance of these driving forces varies greatly with the industry and the situation. Mature industries generally do not have rapidly growing markets and are primarily competing for market share. For these industries, reductions in cost and incremental advantages in perceived or actual performance may represent success (e.g., automobiles). In contrast, technological advances can create large new markets or substantially increase existing markets for newer, rapidly changing industries (e.g., computing). Even when compelling driving forces for change are present, the technological and business risks may be obstacles to commercialization.

Product/Property Data

In the past, primary materials suppliers were only involved peripherally in the design process. As the competition for primary materials has intensified, however, they have become increasingly involved in developing their own design activities. This is especially true for new materials concepts, for which the supplier infrastructure might not be able to meet the needs of industry or for which prospective suppliers may have underestimated the challenges of scaling up an unproven technology. Materials suppliers must collaborate with end-user industries to determine the type of data required for product designers to assess a new material/process and to present the material properties in terms that are relevant and understandable to designers. The committee believes that the precompetitive, cooperative development of product and property data will improve the usefulness of results to product designers. The sharing of basic materials property data might require a review of antitrust legislation and a neutral body (such as the National Institute for Standards and Technology or the American Society for Testing and Materials) as a clearinghouse. Tests and methods should be standardized as much as possible to minimize duplication.

Research Infrastructure

Factors that limit the materials and parts supplier industries as a source of innovation include (1) initial market sizes and profit margins too small to produce adequate return on investment, (2) unwillingness of OEMs to adopt technologies invented by others, and (3) the difficulty in implementing changes to existing supply chains and infrastructure. The research infrastructure for materials and parts supply companies could be improved by the development of mechanisms for larger OEMs to assist and encourage materials supply companies to conduct R&D (e.g., guarantees to use the new technology); government programs, such as ATP, that would help defray some of the costs of industrial R&D; and modifications to the tax code that would permit deductions for R&D expenditures and reduce the risk to the supplier companies.

Patent Protection

If the time required to certify a new material/process approaches the limits of the patent-protection period, a company may not have time to recoup its R&D investment before its competitors can legally use the technology. Because of this, industry tends to be biased toward technologies that can be implemented quickly and leaves more time to accrue profits and recoup R&D investments.

Industrial Ecology

The MS&E community and product designers are increasingly turning to the developing field of industrial ecology to assess the social, economic, and environmental context within which materials and products are designed, produced, used, and managed at the end of their life cycles. This systems-based view of the material includes (1) acquisition; (2) formulation, processing, and manufacturing; (3) distribution as a material or component of a product; (4) use; (5) recycling as part of a refurbished product, assembly, subassembly, component, or material; and (6) eventual disposal or management of the product as waste.

Regulatory Climate

To comply with environmental regulations, industry may have to (1) modify or replace an existing manufacturing process or production facility to reduce harmful emissions or (2) modify or augment a product design to improve safety or reduce harmful emissions. These changes do not generally give any particular company a competitive advantage because they all must comply. In fact, regulations can spur innovations by helping companies bypass the cost barriers for the introduction of new materials/processes and encouraging companies to conduct cooperative, precompetitive research.

Recommendation 4-3. The following developments should be given priority to improve the transition of materials advances from Phase 2 to production implementation:

- collaboration with end-user industries to identify the type of data required for product designers to assess new material/processes (Recommendation 3-1)
- investigation of methods to improve the research infrastructure for materials suppliers and parts suppliers (Recommendations 3-2 and 3-3)
- extension of the patent protection period, especially for applications that require lengthy certification periods (Recommendation 3-4)
- development of industrial ecology as an integral part of the education and expertise of both MS&E researchers and product designers (Recommendation 3-6)
- development of a regulatory climate based on constructive cooperation and goal setting to promote the adoption of new materials that achieve or enhance societal goals (Recommendation 3-16)

References

Allenby, B.R. 1999. Industrial Ecology: Policy Framework and Implementation. Upper Saddle River, N.J.: Prentice Hall.

Allison, J. 1998. Light Weight Connecting Rod: Alternate Materials Case Study. Presented at the Automotive Workshop, National Research Council, Washington, D.C., March 12–13, 1998.

Almasi, G., H. Chang, G.E. Keefe, and D.A. Thompson. U.S. Patent 3691540. Integrated Magneto-Resistive Sensing of Bubble Domain. Filed 10/6/70, Issued 9/12/72.

Baibich, M., J. Broto, and A. Fert. 1988. Giant Magnetoresistance of (001)Fe/(001)Cr magnetic superlattices. Physical Review Letters 61(21): 2472.

Briant, C.L., and B.P. Bewlay. 1995. The Coolidge process for making tungsten ductile: The foundation of incandescent lighting. Materials Research Society Bulletin 20(8): 67–73.

Bridenbaugh, P. 1998. Aluminum-Intensive Vehicles. Presented at the Automotive Workshop, National Research Council, Washington, D.C. March 12–13, 1998.

Buch F. 1998. Aluminum-MMC Disk Brake Rotors. Presented at the Automotive Workshop, National Research Council, Washington, D.C. March 12–13, 1998.

Burte, H.M. 1981. Middle-ground R&D—how can it be rejuvenated? JOM 33(5): 29–30.

Dieny, B, V. Speriosu, and S. Metin. 1991. Magnetotransport properties of magnetically soft spin-valve structures (invited). Journal of Applied Physics 69(8): 4774–4780.

DOC (U.S. Department of Commerce). 1998. Annual Survey of Manufacturers: Statistics for Industry Groups and Industries. M96(AS)-1. Economics and Statistics Administration, Bureau of the Census. Washington, D.C.: U.S. Department of Commerce.

DOC. 1999. Advance-1997 Economic Census: Core Business Statistics Series. EC97X-CS1. Economics and Statistics Administration, Bureau of the Census. Washington, D.C.: U.S. Department of Commerce.

Egelhoff, W.F., Jr., P.J. Chen, C.J. Powell, M.D. Stiles, R.D. McMichael, C.-L. Lin, J.M. Sivertsen, J.H. Judy, K. Takano, A.E. Berkowitz, T.C. Anthony, and J.A. Brug. 1996. Optimizing the giant magnetoresistance of symmetric and bottom spin valves. Journal of Applied Physics 2A 79(8): 5277–5279.

Fisher, J.C., and R.H. Pry. 1971. A simple model of technological change. Technological Forecasting and Social Change 3: 75–88.

Giamei, A.F. 1998. Efficient Materials R&D. Presentation at the Gas Turbine Workshop, National Research Council, Washington, D.C., January 22–23, 1998.

Holton, G., H. Chang, and E. Jurkowitz. 1996. How a scientific discovery is made: a case history. American Scientist 84(4): 364–375.

Hunt, R.P. U.S. Patent 3,493,694. Magnetoresistive Head. Filed 1/19/66, Issued 2/3/70.

Maurer, G.E. 1998. Special Metals Corporation: Alloy/Material Supplier Issues. Presentation at the Gas Turbine Workshop, National Research Council, Washington, D.C., January 22–23, 1998.

McCracken, J. 1998. Tailored Blanks. Presented at the Automotive Workshop, National Research Council, Washington, D.C. March 12–13, 1998.

NAS (National Academy of Sciences). 1998. International Benchmarking of U.S. Materials Science and Engineering Research. Washington, D.C.: National Academy Press.

NRC (National Research Council). 1989. Materials Science and Engineering for the 1990s. Board on Physics and Astronomy and National Materials Advisory Board. Washington, D.C.: National Academy Press.

NRC. 1993. Commercialization of New Materials for a Global Economy. National Materials Advisory Board. NMAB-465. Washington, D.C.: National Academy Press.

NRC. 1996. Coatings for High-Temperature Structural Materials: Trends and Opportunities. National Materials Advisory Board. NMAB-475. Washington, D.C.: National Academy Press.

NRC. 1997. Intermetallic Alloy Development: A Program Evaluation. National Materials Advisory Board. NMAB-487-1. Washington, D.C.: National Academy Press.

NRC. 1999a. Harnessing Science and Technology for America's Economic Future. Washington, D.C.: National Academy Press.

NRC. 1999b. Industrial Technology Assessments: An Evaluation of the Research Program of the Office of Industrial Technologies. National Materials Advisory Board. NMAB-487-4. Washington, D.C.: National Academy Press.

NSTC (National Science and Technology Council). 1995. The Federal Research and Development Program in Materials Science and Technology. Washington, D.C.: U.S. Government Printing Office.

OIT. 1997. The Office of Industrial Technologies: Enhancing the Competitiveness, Efficiency, and Environmental Quality of American Industry through Technology Partnerships. Office of Industrial Technologies, Office of Energy Efficiency and Renewable Energy, U.S. Department of Energy. Washington, D.C.: Department of Energy.

Olson, G.B. 1998. Systems Design of Advanced Alloys: Strategies for Accelerating New Materials Adoption. Presentation at the Gas Turbine Workshop, National Research Council, Washington, D.C., January 22–23, 1998.

Roberge, G.D. 1998. Assessing Performance Potential and Accompanying Risk. Presented at the Gas Turbine Workshop, National Research Council, Washington, D.C., January 22–23, 1998.

APPENDICES

A

Electronics Industry Workshop

O N NOVEMBER 13–14, 1997, the Committee on Materials Science and Engineering: Forging Stronger Links to Users of the National Materials Advisory Board hosted a workshop on linkages and the exchange of information in the electronics industry. This was the first of three workshops intended to identify (1) user needs and business practices that promote or restrict the incorporation of materials and process innovations, (2) how priorities in materials selection are determined, (3) mechanisms to improve links between the materials community and the engineering disciplines, and (4) programs (e.g., education, procedures, information technology) to improve these linkages.

As shown in the agenda in Box A-1, the workshop was divided into four sessions. The first three sessions were devoted to different aspects of the electronics industry: the magnetic hard-disk-drive (HDD) industry, the chip manufacturing industry, and the packaging industry. Each session included presentations by representatives of consortia, academia, industrial research and development (R&D) organizations, supply manufacturers, and primary manufacturers. The fourth session was devoted to a discussion of the characteristics of the electronics industry that distinguish it from other industries and the importance of road maps and linkages among primary industries, supplier industries, and universities in the development of advanced technologies.

MAGNETIC HARD-DISK DRIVE INDUSTRY

The HDD market in 1997 was estimated to be $35 billion. Sixty-one million HDDs were shipped worldwide in the first half of 1997. The companies involved were Seagate (24.5 percent), Quantum (20.4 percent), Western Digital (19.7 percent), IBM (11.7 percent), Fujitsu (7.7 percent), Maxtor (4.9 percent),

BOX A-1
Agenda for the Electronics Industry Workshop

November 13, 1997
8:30 a.m.   Convene and Introductions, Dale F. Stein, Committee Chair

MAGNETIC STORAGE SESSION (W. Doyle, Session Chair)
9:00 a.m.   Road Map Development and Maintenance, B. Schechtman, *NSIC*
9:20 a.m.   Source of Invention: University, Sheldon Schultz, *UCSD*
9:40 a.m.   Source of Invention: Industry, David Thompson, *IBM Almaden Research Center*
10:00 a.m.   Supply Industry Perspective, R. Rottmayer, *Read-Rite*
10:20 a.m.   User Industry Perspective, Thomas Howell, *Quantum*
10:40 a.m.   Discussion

CHIP MANUFACTURING SESSION (J. Shaw, Session Chair)
1:00 p.m.   Road Map Development and Maintenance, Paul Peercy, *SEMI/ SEMATECH*
1:20 p.m.   Source of Invention: University, Woodward Yang, *Harvard University*
1:40 p.m.   Source of Invention: Industry, Don W. Shaw, *Texas Instruments*
2:00 p.m.   Supply Industry Perspective, Alain Harrus, *Novellus*
2:20 p.m.   User Industry Perspective, Pier Chu, *Motorola*
2:40 p.m.   Discussion
5:00 p.m.   Adjournment

November 14, 1997
8:30 a.m.   Convene, Dale F. Stein, Committee Chair

PACKAGING SESSION (J. Decaire, Session Chair)
8:35 a.m.   Road Map Development and Maintenance, James McElroy, *NEMI*
8:55 a.m.   Source of Invention: University, Michael G. Pecht, *University of Maryland*
9:15 a.m.   Source of Invention: Industry, William T. Chen, *IBM*
9:35 a.m.   Supply Industry Perspective, Jack Fischer, *Interconnection Technology Research Institute*
9:55 a.m.   User Industry Perspective, Robert MacDonald, *Intel*
10:15 a.m.   Discussion

DISCUSSION SESSION
1:00 p.m.   Generic Linkages in the Electronics Industries
2:00 p.m.   Strengths and Weaknesses of Linkages in the Electronics Industries
3:00 p.m.   Strategies for Improving Linkages in the Electronics Industries

Toshiba (4.7 percent), and others (6.4 percent). Although drive design is still concentrated in the United States, more than 50 percent of head and media development and manufacturing occurs elsewhere. Most HDDs are assembled in the Far East (50 percent in Singapore).

The basic components of an advanced HDD system are (1) moving magnetic media with two remnant states (representing "1" and "0" bits in digital systems); (2) a magnetic head with a miniature transformer for recording and magnetoresistive (MR) field sensors for reading; (3) an interface between the head and media to achieve high reliability at extremely small spacings (25 nm); (4) a servo system for tracking previously written data; and (5) electronics to detect data at an error rate of less than 10^{-12} errors/bit. The first three components—which include disk substrates and surface overcoats, magnetic film media, wear-resistant overcoats, topical lubricants, head carriers (sliders), magnetic films for record heads, complex multilayer MR structures for reading, conductors, insulators, and planarization materials—are materials intensive and were the focus of the HDD session of the workshop.

HDD industry characteristics were identified as follows: (1) high volume at low cost; (2) short product cycles (less than 2 years); (3) low profit margin; (4) successful incremental improvements; (5) complex supplier networks; and (6) proprietary manufacturing process "art."

The current metrics for evaluating HDD systems are cost, capacity, and access time. Incredible progress has been made in capacity, driven by the areal storage density, which has increased six-fold in the past 50 years. This progress has been driven primarily by scaling dimensions, which required significant changes in materials and processes. In the last 15 years, six major changes have been implemented in media (i.e., thin-film media and glass substrates) and heads (i.e., metal-in-gap, thin-film inductive, thin-film MR, and thin-film giant MR).

Until the early 1980s, almost all HDD technology was originated by IBM. Since then, as competition has increased and IBM has scaled back its R&D, the responsibility for the development of new materials and processes has fallen more to other manufacturers. Sources of information on new materials include industry alliances, mergers, technical conferences, publications, and contract research with university faculty members.

In the 1980s, only a few university faculty members were interested in magnetic devices, so most new graduates were trained by industry. Two substantial changes have improved the linkages between universities and the HDD industry since then:

- the establishment of industry-supported, multidisciplinary university centers devoted to magnetic storage (first at Carnegie-Mellon University and the University of California at San Diego, and later at the University of Alabama, the University of Minnesota, the University of Washington, the University of

California at Berkeley, Ohio State University, the University of Nebraska, and Rice University)
- the formation in 1990 of the National Storage Industry Consortium (NSIC) to provide mechanisms for industry-university collaborations on focused problem areas

NSIC's mission is to increase the worldwide competitiveness of the U.S. storage industry by: (1) conducting joint research on high-risk, precompetitive storage technologies; (2) procuring government funding; (3) developing technology road maps; (4) maximizing the value of university research; and (5) acting as an industry spokesgroup.

The following characteristics of the HDD industry were identified by individual workshop participants as contributors to the introduction of advances in materials and processes:

- competition in an industry that produces high-technology products
- extraordinary improvements in performance fueled by materials innovations
- extensive industry-university collaborations facilitated by NSIC
- strong university centers focusing on storage technologies, often staffed by faculty with industry experience
- NSIC-developed technology road maps that identify challenges for continued progress
- no regulatory issues to interfere with efforts to develop new technology
- high employee mobility, which provides rapid equilibration of technology

The following characteristics of the HDD industry were identified by workshop participants as inhibiting the introduction of materials and process advances:

- strong interdependence of heads, media, interface, servo, and channel electronics, which makes it difficult to make changes that could affect more than one component
- low profit margins in a commodity market, which shifts the emphasis to evolutionary rather than revolutionary changes
- complex intellectual property agreements that inhibit universities from obtaining patent protection
- inconsistencies between university "blue sky" research and focused industry goals
- insufficient federal support for the mainstream magnetic storage industry
- requirements for extensive empirical investigations because materials modeling capabilities are insufficient
- ineffective accelerated tests for the evaluation of long-term reliability

- requirements for large capital investments ($250 million/year) 12 to 18 months in advance of orders
- shrinking customer base for materials suppliers

CHIP MANUFACTURING INDUSTRY

The second session was devoted to the chip manufacturing industry. The increased density of silicon, with feature sizes being reduced at a rate of about 10 percent per year, will lead to a predicted market value of $200 billion by the year 2000. Complementary metal oxide on silicon (CMOS) technology dominates more than 90 percent of the market. The dimensions of current CMOS chips are approximately 2 cm × 2 cm, with 0.35 micron critical feature sizes and three to five layers of metal wiring to interconnect the devices. The chips are fabricated on an 8-inch wafer, which requires about one month of processing time (three to six steps/day, running 24 hours/day). Terabucks (on the order of 10^{12}) are currently invested in infrastructure, including raw materials, equipment, and R&D.

Since 1992, the Semiconductor Industry Association has coordinated a process to develop a road map of industry technology requirements with a 15-year horizon. The market has grown at a rate of 15 percent per year for the past 35 years, following Gordon Moore's prediction of a 20 to 25 percent per year improvement in cost performance through (1) shrinking feature sizes (which increases performance), (2) increasing wafer sizes, (3) improving yield, and (4) increasing manufacturing productivity (which lowers costs). As the industry moves into the production of feature sizes of 0.1 micron within the next 10 years, however, innovative technologies will have to be developed. The National Technology Roadmap for Semiconductors, which identifies the key technology needs, was devised by a large cross-section of the semiconductor community. As many as 600 engineers from industry, government, universities, and suppliers participated in the technical working groups (TWGs), which included lithography, interconnects, front-end processes, factory integration, assembly and packaging, design and testing, process integration, devices, and structures. Crosscutting TWGs focused on environment, safety, and health; metrology; defect reduction; and modeling and simulation. The 1997 road map identified six difficult challenges facing the semiconductor industry that will require major initiatives to overcome: (1) continued affordable scaling; (2) affordable lithography at and below 100 nm; (3) on-off chips that operate at GHz frequencies; (4) new materials and structures; (5) measurements, metrology, and testing; and (6) R&D challenges.

New materials are being explored at all levels for future silicon chips, from new substrate materials to new gate and gate oxide systems to high dielectric constant electrode materials to low dielectric constant insulators for wiring interconnections. The incentives to develop these materials are performance enhancement, product needs, cost reduction, competitive advantage, and regulatory issues. The obstacles to material implementation are manufacturability and cost, high risk, long development times, lack of control on tool development (long lead time), cross-contamination in fabrication, difficulty in predicting time-to-market, and inadequate information on critical materials.

Suppliers of manufacturing equipment do not dictate materials choices, but they enable their use in manufacturing. The mean time from development of a new material to manufacturing implementation is six to seven years, and the time to develop processing equipment increases this to ten to fifteen years. These long development times reflect the extreme conservatism of the semiconductor industry, which prefers to make progress through careful evolutionary tweaking. Major changes that are considered revolutionary are implemented only when absolutely necessary (e.g., when mandated by performance, regulation, or cost requirements).

From the industry perspective, some of the risk and cost can be reduced by leveraging university research and ideas to complement internal R&D and develop science and technology for future products. Although communication has been improved in recent years, there are still a number of weak links between the industry and universities: differences in culture and objectives (e.g., system solutions and manufacturability are not usually objectives for universities); differences in policies and practices on intellectual property; and the lack of material and simulation/prediction techniques. These links could be improved in several ways: addressing integrated-product issues early; promoting personnel exchanges/long-term visits; clearly defining objectives and milestones; providing adequate project reviews and industry mentors; and strengthening the industrial/university research community through participation in consortia and jointly sponsored conferences.

The development of material technology, which is essential to the microelectronics industry, is a high-risk, high-reward undertaking that requires vision and a long time span. University research is an important part of industrial materials technology development and can reduce the risk and costs to industry. The implementation of new materials in products could be accelerated by (1) leveraging strategic programs to provide early learning and materials/modeling predictions; (2) closely adhering to product-technology road maps, (3) focusing on manufacturability and cost issues early; and (4) developing a systems approach to materials development.

The following factors were identified by workshop participants as contributing to the introduction of advances in materials and processes:

- Road maps provide a technological "stake in the ground" so that chip manufacturers can focus on accelerating time-to-market.
- Road maps provide a framework for industry, suppliers, academia, and

government to "buy into" goals and research directions and provide a tool to help funding agencies decide which projects to fund.

- Road maps and short product cycles allow universities to focus on long-term goals to extend the fundamental limits of silicon devices.
- Industrial laboratories can focus on solving critical near-term problems.
- Equipment suppliers can develop hardware from concept stage to production tools in approximately four years.

The following factors were identified by workshop participants as inhibiting the introduction of materials and process advances:

- Technology road maps, even if they take a long-term view, are essentially evolutionary.
- Production conditions are difficult to duplicate in a university research environment because of the costly tooling or processing of real microelectronics fabrication facilities.
- Increasing standardization places bounds on materials innovation.
- Availability of equipment and processes are setting the pace of materials innovation.
- Equipment manufacturers depend on the rapid development and acceptance of new materials for the timely development of new tools.
- Equipment manufacturers take a substantial risk in the introduction of new materials because there is no infrastructure available to test integration in a manufacturing line.

PACKAGING INDUSTRY

The third session was devoted to a discussion of the packaging industry. Microelectronic packaging can be considered at several levels: silicon wafer/chips, chip carriers, printed wiring boards and other interconnection substrates, circuit card assembly/test, and final product assembly/test. At one end, packaging technology is being driven by advances in silicon technology and at the other end by customer acceptance in the marketplace (e.g., form, fit, function, cost).

Trends in electronics packaging technologies were illustrated through selected examples. The most advanced packaging technologies are being implemented in portable hand-held products, such as the Sony camcorder and the Motorola cellular phone. These products are exploiting small, lightweight chip carriers with high input/output (I/O) pin densities, such as chip-scale packages and ball grid arrays as well as flip-chip and chip-on-board assembly technologies. Trends in packaging for integrated circuits included the following:

- ball grid array and chip-scale packaging
- area-array chip interconnections

- migration from ceramic to organic materials
- high-density substrates for ball-grid array assemblies
- multichip packaging
- package design for area/volume and weight constraints

New materials are generally introduced one at a time to minimize requirements for new tooling and to reduce capital investment and materials risks. The key decision factors are low technical risks, readily available material sources, and low costs.

The National Electronics Manufacturing Initiative (NEMI) is a private, industry-led consortium dedicated to the advancement of the electronics manufacturing infrastructure in North America. NEMI members include a broad spectrum of electronic equipment manufacturers, components suppliers, manufacturing equipment suppliers, contract manufacturers, material suppliers, software suppliers, consortia/ trade associations/consultants, universities, and government organizations.

NEMI's road map structure is derived from a comprehensive manufacturing system plan (i.e., a manufacturing system model representative of a comprehensive supply chain and factory for a "virtual product" target anticipated to be representative of future high-volume products). NEMI has customized the generic target product according to different market sector drivers. Five product market sectors are characterized as follows: low cost, hand held, cost/ performance, high performance, and harsh environment. Manufacturing infrastructure needs for each element of the plan were assessed to develop road maps. The overall NEMI road map is a coordinated set of road maps for each of the following manufacturing system elements: packaging; board (circuit card) assembly; final product assembly; interconnection substrates; displays; energy storage systems; radio-frequency components; passive components; semiconductor devices; magnetic mass data storage; optical mass data storage; optoelectronics; factory information systems; modeling, simulation, and rapid prototyping; and test, inspection, and measurement. NEMI has developed interorganizational linkages to coordinate these road maps (e.g., magnetic and optical storage; semiconductor devices; displays; optoelectronics; and interconnection substrates).

The manufacturers of electronic end products have embraced technology road maps as a way to establish linkages with their suppliers, including the R&D community. The NEMI road map does not explicitly address materials as a separate element of the manufacturing system plan, but materials issues are embedded in each element. Some workshop participants were concerned that the road mapping process tends to emphasize evolutionary development and thus may overlook revolutionary innovations.

Several mechanisms for research collaboration between industry and universities are being implemented. University centers with research agendas based on problems identified by industrial partners are attracting more industrial participants and funding. Presumably, their research results will be more rapidly adopted.

FINAL DISCUSSION

The final afternoon of the workshop was devoted to a discussion of (1) the characteristics that distinguish the electronics industry from other industries and (2) the importance of road maps and linkages among primary industries, supplier industries, and universities in the development of advanced technologies. Workshop participants identified two characteristics that distinguish the electronics industry from the automotive and turbine-engine industries:

- The motivation to improve product performance has been based more on internal industrial road mapping (i.e., Moore's Law) than on customer relations (i.e., demands of the buyers and users of computers).
- Because of the emphasis on performance qualification, developers of electronic components have greater freedom to incorporate new materials and processes.

Some participants questioned whether the electronics industry actually introduced new materials faster than other industries. The basic materials and processes used in the electronics industry have remained relatively constant for the past 20 to 30 years. Silicon technology has improved incrementally, which has enabled the reduction of component sizes and improvements in performance.

Many workshop participants felt that road maps are especially important for complex products. They identified four ways in which road maps could further technological development:

- by defining the issues facing industries and gaps in technology
- by coordinating all segments of the industry in the development process—from fundamental research and development to final assembly
- by providing a framework for industries and researchers to plan potential materials changes and technological innovations
- by providing a level playing field among researchers and industries and lowering overall risk

Consortia play two important roles in the road-mapping process: (1) providing neutral territory on which competing industries can meet to develop and maintain industrial road maps and (2) acting as links between industries and universities to ensure that the necessary short-term research is conducted. In the opinion of some participants, consortia and industry have been too effective in promoting short-term research at universities, to the detriment of long-range research and teaching.

Individual workshop participants identified three weaknesses in the industry-university collaboration:

- Many primary manufacturing companies support university research, but supply industries have not been as active. This linkage could be strengthened by the development of more supplier-oriented road maps.
- Most universities conduct research on different equipment than the equipment used in industry, which reduces the compatibility and ultimate usefulness of some research results. If research at universities leads to the development of better process models, and thus a better general understanding of the fundamental principles of new systems, industrial parameters could be determined and the transition of new materials and processes accelerated.
- Many universities are devoting considerable time and resources to the establishment of university-industry links. Universities and industries should develop a standard methodology for interactions to eliminate the need to reinvent contracts with each new project.

Most participants felt that the linkages between the primary and supply companies were stronger in the magnetic-head industry and the chip-manufacturing industry than in the packaging industry. An enormous amount of information is exchanged between all electronics industries to ensure that suppliers' products meet the needs of the primary manufacturers. One of the strengths of this relationship has been the standardization of many features (e.g., inputs, outputs, performance indicators). The participants noted that SEMI/SEMATECH has been instrumental in the establishment of this close integration.

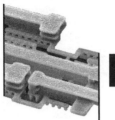

B

Turbine-Engine Industry Workshop

O N JANUARY 22–23, 1998, the Committee on Materials Science and Engineering: Forging Stronger Links to Users of the National Materials Advisory Board hosted a workshop on the linkages and exchange of information in the turbine-engine industry. This was the second of three workshops intended to identify (1) user needs and business practices that promote or restrict the incorporation of materials and processes innovation, (2) how priorities in materials selection are determined, (3) mechanisms to improve links between the materials community and the engineering disciplines, and (4) programs (e.g., education, procedures, information technology) to improve these linkages.

As shown in the agenda in Box B-1, the workshop was divided into four sessions. The first two sessions were devoted to different sectors of the turbine-engine industry—primary manufacturing and supply industries. The third session was devoted to business issues. The fourth session was devoted to a discussion of the distinguishing characteristics of the turbine-engine industry and the linkages among primary industries, supplier industries, and universities in the development of advanced technologies.

Aircraft turbine engines are the single largest U.S. export product. The industry serves both commercial and military customers, whose missions, needs, and priorities are very different. The balance of effort between these two classes of customers is subject to business cycles. In terms of materials technology, this industry is closely linked to the turbine power-generation industry. Although the aircraft turbine business is closely regulated by the Federal Aviation Administration, engine producers and the regulatory agency enjoy a productive relationship because they share goals with respect to flight safety.

At first glance, the jet engines industry appears to be a conventional materials supply chain involving raw materials suppliers, value-added distributors, parts

BOX B-1
Agenda for the Turbine Engine Industry Workshop

January 22, 1998

8:30 a.m.	Convene and Introductions, Dale F. Stein, Committee Chair
8:50 a.m.	Overview from the Design Perspective, Ambrose Hauser, *GE Aircraft Engines*
9:10 a.m.	Technology Acquisition and Insertion, Michael Goulette, *Rolls-Royce PLC*
9:30 a.m.	Overview of Performance Risks, Gary Roberge, *Pratt and Whitney*

PRIMARY MANUFACTURING SESSION (Malcolm Thomas, Session Chair)

9:50 a.m.	Historical Case Study: Single-Crystal Blades, Anthony Giamei, *United Technologies*
10:10 a.m.	Current Case Study: Metal Matrix Composites, Kathy Stevens, *Wright Laboratories*
10:30 a.m.	Current Case Study: Titanium Aluminides, James Williams, *GE Aircraft Engines*
10:50 a.m.	Integration of Materials with Design Requirements, Peter Shilke, *GE Corporation*
11:10 a.m.	Discussion

SUPPLY INDUSTRIES SESSION (Neil Paton, Session Chair)

1:20 p.m.	Historical Case Study: Thermal Barrier Coatings, Harry Brill-Edwards, *consultant*
1:40 p.m.	Disk Process Modeling, Robert Noel, *Ladisch Company, Incorporated*
2:00 p.m.	Alloys Design, Greg Olson, *QuestTek Innovations*
2:20 p.m.	Strategies to Reduce Cycle Times and Hit Opportunity Windows, Gernant Maurer, *Special Metals Corporation*
2:40 p.m.	Discussion
5:20 p.m.	Adjournment

January 23, 1998

BUSINESS ISSUES SESSION (William Manly, Session Chair)

8:35 a.m.	Business Aspects, Ken Harris, *Cannon-Muskegon Corporation*
8:55 a.m.	Road maps: Advanced Turbine Systems, William Parks, *U.S. Department of Energy*
9:15 a.m.	FAA Regulatory Issues, Mark Fulmer, *Federal Aviation Administration*
9:35 a.m.	Liability and Regulatory Issues, Tony Freck, *consultant*
9:55 a.m.	DARPA Insertion Program, Larry Fernbacher, *Technology Assessment and Transfer*
10:15 a.m.	Discussion

DISCUSSION SESSION

1:00 p.m.	Generic Linkages in the Jet Engine Industry
2:00 p.m.	Strengths and Weaknesses of Linkages in the Jet Engine Industry
3:00 p.m.	Strategies for Improving Linkages in the Jet engine Industry

makers, and original equipment manufacturers (OEMs), all of whom are involved to varying degrees in the process of developing and commercializing new materials. This rather conventional structure belies many unique features of the jet engines business that can only be seen by examining the links and interactions among supply chain participants.

At one end of the chain are the raw materials suppliers, mainly mining and metal-refining companies involved in the production of nickel, titanium, and a host of alloying elements ranging from aluminum to zirconium. Typical jet engine alloys are made up of 10 or more metallic elements, most of which are supplied by independent companies. Generally, these companies supply their products to many industries for uses other than jet engines and, therefore, are not commercially dependent on the jet engine business for their livelihood.

Because jet engine alloys are a complex and carefully controlled mixture of many elements, the supply chain role also includes "mixologists," in this case specialty metal suppliers who melt and mix the ingredients of jet-engine alloys and also perform a host of other value-added activities to ensure the quality and integrity of the alloys.

The third major link in the jet-engine supply chain is of the parts maker. Companies in this position generally focus on a particular manufacturing technology (e.g., casting, forging, or machining). They convert the metal alloys produced by specialty metal suppliers into finished components for installation into jet engines. One simple, but significant, characteristic of parts makers is that they buy by the pound and sell by the piece. Parts makers sell components to the jet engine manufacturer, or, more accurately, they are contracted by the engine manufacturer to produce components. The engine manufacturer inspects the parts, accepts or rejects them, and incorporates them into the engines.

Engine producers are not at the end of the supply chain but are a step closer to the end of the chain than is generally assumed. In the commercial aviation market, jet engines are often sold directly to the final customer—the commercial airline company—rather than to the airframe manufacturer. Thus, Delta Airlines rather than Boeing is likely to be the jet engine producer's customer.

Although the structure of this supply chain is unremarkable, the nature of the interactions among the members is unique. One unique feature of the jet-engine supply chain is that large parts of it are made up of technological oligopolies. For instance, three or fewer producers of superalloy, producers of titanium, forgers, and foundries service the entire industry. These oligopolies combined are responsible for producing a significant fraction of the technological content and the majority of the weight of an engine. Unlike the raw materials suppliers, these companies are almost entirely dependent on the jet-engine business for their livelihood. Consequently, whereas it might be assumed that they would enjoy certain oligopoly privileges and be able to extract excess profits, there is little evidence of this occurring.

NEW MATERIALS DEVELOPMENT: INCENTIVES AND BARRIERS

In the discussion of driving forces for materials selection and the implementation of new materials technology, workshop participants pointed out that the OEM's design organizations ultimately select materials on the basis of potential performance enhancements and customer requirements for cost benefits. Many workshop participants felt that the industry is undergoing an uneasy shift from performance-based materials selection processes to processes based on both cost and performance. Because customers include military aviation, commercial aviation, and electric utilities, performance and cost needs vary and change frequently, which makes it difficult for materials developers to establish consistent research objectives.

The jet-engine industry does not have an obvious technology strategy (road map) for developing commercially driven, next-generation products. Engine manufacturers do not disclose their thoughts on future material needs to suppliers because the structure of the industry and the relationships with end-users makes it difficult to limit the diffusion of a new technology long enough for an innovator to sustain a technical competitive advantage from materials technologies. Thus, even though engine manufacturers have proprietary road maps, there is no agreement on industry-wide development goals. The Department of Energy's Advanced Turbine Systems and National Aeronautics and Space Administration's High Speed Civil Transport programs have come closest to developing program road maps, probably because they were designed as precompetitive technology development programs.

Workshop participants identified three characteristics of the jet-engine industry that encourage the development and implementation of new materials:

- The industry is aware that many improvements in engine performance have resulted from improvements in materials capabilities. For example, stringent performance requirements led to the development and introduction of multiple generations of wrought and cast nickel-base alloys with increasingly higher temperature capabilities.
- The industry is convinced that improvements in materials can significantly enhance customer-driven engine performance metrics.
- The industry recognizes the potential payoffs of future materials improvements. For example, improvements could decrease specific fuel consumption, reduce component or system weight, increase thrust-to-airflow ratio, and/or improve the durability and reliability of jet-engine systems.

The workshop participants then identified business and technological factors that acted as barriers to the development and introduction of new materials:

- The industry was established as a research and development (R&D) industry that has relied heavily on government funding for the development of leapfrog technologies. As a result, the industry has historically pursued R&D on high-risk materials and processes, which have significantly improved both military and commercial aircraft engines. Despite the historical focus on high-risk technologies, the industry is extremely cautious about integrating new materials into service because of cost and/or the risk of failures.

- Manufacturers and end users both require a lengthy, expensive qualification cycle to develop a thorough characterization of the performance, durability, and reliability of new materials before they can be considered for insertion in engines.
- The cost of development rises steeply as a prelaunch effort progresses from basic research to the demonstration of full-sized components.
- The time required to complete characterization and qualification phases is longer than the engine design cycle. As a result, support for technology development is generally inconsistent and discontinuous.
- Because of the complexity of engine systems, it is difficult to assess the potential impact of new materials on performance.
- Poor communications between mechanical designers and materials developers have resulted in "missed opportunities" for the introduction of new materials.
- New materials often originate in the engine manufacturer's laboratories, which discourages suppliers from developing new materials.
- Requirements for multiple sources of materials and the absence of alternative markets are disincentives for materials suppliers to develop new materials/processes.
- The profit margin for engine manufacturers is narrow because their recently deregulated, financially sophisticated customers (airlines) have struggled to maintain their own profitability.

IMPROVING LINKAGES

Workshop participants suggested that the following steps be taken to improve linkages between the MS&E community and engine manufacturers:

- Maintain consistent funding throughout the development cycle, from research through insertion.
- Establish collaborative precompetitive programs, with suppliers and engine manufacturers working in teams on critical materials technologies for more directed explorations of the transition from research to development.
- Maintain significant industry involvement in university research programs, and target specific gaps in knowledge with regard to new materials and processes.

- Provide standardize test methods and specifications to suppliers.
- Focus materials research on near-term, incremental, cost-effective technologies.
- De-emphasize high-risk, leapfrog technologies.
- Establish general industry guidelines, based on an open exchange of information among suppliers and engine builders, that state specific needs and goals in terms of materials and processes.
- Make greater use of computer modeling (e.g., process models, thermodynamic and kinetic models of structural development, life prediction models) to reduce the cost, risk, and time involved in materials development.
- Put more emphasis on business metrics (especially manufacturing cost and life-cycle cost analyses) in selecting and evaluating R&D programs.
- Involve certification authorities early in the technology development cycle.
- Make use of consortia and university centers of excellence for precompetitive programs.
- Establish a forum for engine designers, material suppliers, and parts suppliers to reach a consensus on the probability and economic viability of advanced materials for the jet engine of tomorrow.

C

Automotive Industry Workshop

O N MARCH 12–13, 1998, the Committee on Materials Science and Engineering: Forging Stronger Links to Users of the National Materials Advisory Board hosted a workshop on the linkages and exchange of information within the automotive industry. This was the third of three workshops intended to identify (1) user needs and business practices that promote or restrict the incorporation of materials and processes innovation, (2) how priorities in materials selection are determined, (3) mechanisms to improve links between the materials community and the engineering disciplines, and (4) programs (e.g., education, procedures, information technology) to improve these linkages. As shown in the agenda in Box C-1, the workshop was divided into four sessions: material selection processes, supplier perspective on alternate materials, and two sessions on alternate materials case studies.

NEW MATERIALS DEVELOPMENT:
INCENTIVES AND BARRIERS

Automotive products are mature but will require innovative alternative materials to continue to compete in the global marketplace, and equally important, to meet future societal and regulatory demands. The industry established a record of responding to these driving forces during the past two decades. For example, from 1975 to 1983 the average vehicle weight was reduced across the entire fleet by 1,200 lbs.

The mix of materials used in automobiles has changed substantially in the past two decades: high-strength steels have increased from 0 percent to 11 percent; cast aluminum from 2 percent to 6 percent; engineering plastics from 0 percent to 10 percent; and mild steel/cast iron has decreased from 75 percent to

---

BOX C-1
Agenda for the Automotive Industry Workshop

March 12, 1998

8:20 a.m.   Convene and Introductions, Dale F. Stein, Committee Chair

OVERVIEW: MATERIAL SELECTION PROCESSES (Ronald Shriver, Session
                Chair)

8:30 a.m.   Ford System, C. L. Magee, *Ford*
9:15 a.m.   GM System, R. Heimbuch, *General Motors*
10:00 a.m.   Introduction of New Materials into Manufacturing Operations, S.
                 Harpest, *Honda*
10:45 a.m.   PNGV Materials Road Map, A. Sherman, *Ford*   .
11:30 a.m.   Discussion

SUPPLIERS PERSPECTIVE ON ALTERNATIVE MATERIALS (N. Gjostein,
                Session Chair)

1:00 p.m.   Optimized Steel Vehicles, D. Martin, *AISI International*
1:45 p.m.   Aluminum Intensive Vehicles, P. Bridenbaugh, *ALCOA*
2:30 p.m.   Composite Intensive Vehicles, K. Rusch, *Budd Plastics*
3:15 p.m.   Panel Discussion: Engineering Plastic Components, K. Browall, *GE*
4:15 p.m.   Discussion
5:00 p.m.   Adjourn

March 13, 1998
ALTERNATE MATERIAL R&D CASE STUDIES I (R. Wagoner, Session Chair)

8:15 a.m.   Tailored Blanks, J. McCracken, *TWB, Inc.*
9:00 a.m.   Alternate Materials for Con-Rods, J. Allison, *Ford*
9:45 a.m.   Aluminum-MMC Disk Brake Rotors; F. Buch, *DURALCAN*
10:30 a.m.   III-V Compound Position Sensors, J. Heremans, *GM*
11:15 a.m.   Titanium Applications, S. Froes, *University of Idaho*

ALTERNATE MATERIAL R&D CASE STUDIES II (J. Busch, Session Chair)

12:45 p.m.   Applications of Structural Ceramics, B. McEntire, *Norton*
1:30 p.m.   Steel vs. Aluminum vs. Polymers in Auto Body Applications,
                 J. Dieffenbach, *IBIS Associates*
2:15 p.m.   Discussion: Strengths and Weaknesses of Linkages in the
                 Automotive Industry
3:15 p.m.   Discussion: Strategies for Improving Linkages in the Automotive
                 Industry

---

57 percent. The use of stainless steel, magnesium, powder-metal parts, zinc-coated body sheet, ceramic honeycombs and sensors, Pt-Rh three-way catalysts, micromachined silicon capacitive pressure sensors, and cathodic electrocoating has also increased.

Workshop participants identified the following strengths of the automotive industry that facilitate the introduction of new materials/processes:

- highly sophisticated, computer-based design techniques (e.g., computer-aided design, computer-aided engineering, and computer-integrated manufacturing) to optimize new material concepts
- rapid prototyping techniques (current and developmental) that can greatly accelerate the introduction of new material concepts
- a huge capital investment in testing facilities
- a large supplier base that works jointly with original equipment manufacturers (OEMs) to develop new material concepts
- a talented engineering workforce with strong materials capabilities
- established links and joint programs with national laboratories and universities
- industry consortia (e.g., the U.S. Automotive Materials Partnership) to establish a materials research and development (R&D) agenda
- a federal program, Partnership for a New Generation of Vehicles (PNGV), which is developing a materials R&D road map

Workshop participants identified the following characteristics of the automotive industry as barriers to the introduction of new materials/processes:

- a short (three to four year) product development cycle that provides regular, but still somewhat limited, opportunities for the insertion of new technologies
- a large, established capital equipment base that is renewed only periodically, which tends to inhibit the adoption of new technologies
- a large existing base of knowledge in conventional materials that tends to promote the status quo
- difficulty in predicting the perceived value of a new technology through cost/benefit analyses
- a risk-averse design community that is leery of introducing new concepts
- a rigid purchasing system that is skeptical of suppliers who do not have a track record of supplying high-quality parts in high volume

IMPROVING LINKAGES

The workshop participants considered many sources of new materials technology, including universities, government laboratories, joint projects with government support (e.g., cooperative research and development agreements [CRADAs], Advanced Technology Program [ATP] initiatives), small entrepreneurial firms, primary material suppliers, parts fabricators, subsystem suppliers/full service suppliers, and OEM R&D laboratories. Many linkages are possible and the paths from the source of a new technology to implementation and commercial success are intimately involved with the product development and manufacturing process, which takes, on average, about three years. Efforts are under way to reduce the cycle time to closer to two years. Even with the present development cycle technology must be virtually fixed at the design phase. Thus, the validation of a new technology must be completed before the decision is made to start a vehicle program because opportunities to develop concepts during the program are few. Assuming that (1) materials development takes three to five years, (2) component testing takes one to two years, (3) manufacturing scale-up takes one to two years, and (4) product cycle time is two to three years, the total cycle takes between 7 and 12 years.

Universities and Government Laboratories

Over the past decade, both universities and government laboratories have been more willing to work on concepts that are more relevant to industry. The major barrier in this linkage is the lack of a supplier infrastructure to supply highly reliable parts in high volume. OEMs generally do not consider themselves as developers of supplier infrastructures for new materials technology but prefer to wait until the technology and supplier infrastructure has been developed for other products. For example, the increase in components on passenger cars and trucks built from engineering plastic components (e.g., interior/exterior trim and assorted small parts) happened in this manner. The stakeholders involved in new materials developments must also try to develop a supplier infrastructure as the technology develops.

Parts Suppliers

Many workshop participants felt that the linkages between auto manufacturers and parts suppliers are very strong and, probably, the most important links. OEMs urge their lower tier suppliers to conduct R&D, either on incremental improvements to existing products or on riskier new concepts.

The strongest linkages are between design and engineering activities by OEMs and corresponding activities in supplier organizations. Linkages between OEM R&D activities and suppliers' R&D have been weak. Suppliers are often reluctant to conduct joint R&D projects with OEMs for a number reasons, mostly related to proprietary restrictions on research results. This situation is changing, however, as PNGV, CRADAs, and ATP initiatives are encouraging precompetitive joint R&D.

Primary Materials Suppliers

Primary materials suppliers of materials, such as steel, aluminum, and plastic resins, serve both parts fabricators (at all tiers) and OEMs. In the past, primary materials suppliers were not active participants in the design process. As the competition for the predominant automobile body material intensifies (e.g., steel, aluminum, or plastic composites), suppliers of these materials have developed their own design activities and have indicated that they want to be involved in the OEMs' product development and design process.

FINAL DISCUSSION

The workshop participants agreed that materials and processing research should focus on areas that will lead to lower emissions, lower cost, greater efficiency, and better fuel-cell options. The PNGV program was cited as a unique industry/government partnership working toward dramatic reductions in vehicle weight and increases in performance. Many participants agreed that extensive use of lightweight materials and other advanced materials and process technologies throughout the industry will be necessary to achieve aggressive goals like those set for PNGV. Through programs like PNGV, lightweight materials could be made more attractive for high-volume automotive applications.

Workshop participants identified the following factors as controlling the decision to implement a new materials technology in the automotive industry:

- cost compared to the existing part or subassembly, including materials, processing, tooling and facilities, and offsets for benefits realized in other subsystems
- high-volume manufacturing process capability
- assurance that the quality, reliability, and durability will be greater than or equal to the existing system
- availability of a supplier infrastructure that can meet the standards of automotive purchasing organizations

Workshop participants agreed that only cost-effective and well proven concepts will be integrated into vehicle programs. Several workshop participants suggested that, even if the cost comparisons are unfavorable, new technology might still be implemented under the following conditions:

- The new technology is a saleable customer feature that can be priced to maintain or increase profits (this is rare for materials concepts).
- The new technology has a favorable effect on warranty that can be calculated from current warranty costs.
- The new technology is required to meet regulations (in this case, the innovation may or may not be recovered in the vehicle price).
- The new technology is required to compete with other producers, and the variable cost increase can be offset either in the same subsystem or by reducing costs in other parts of the vehicle.
- The new technology helps to overcome the "guzzler tax" and considers variable cost, publicity, and effect on market share.

D

Biographical Sketches of Committee Members

DALE F. STEIN (chair) retired from his position as professor of materials science and engineering at Michigan Technological University and is now president emeritus of the university. He has also held positions at the University of Minnesota and the General Electric Research Laboratory. He is an internationally renowned authority on the mechanical properties of engineering materials and has served on numerous advisory committees for the National Science Foundation (NSF), the U.S. Department of Energy (DOE), and the National Research Council (NRC). Dr. Stein received the Hardy Gold Medal of the American Institute of Mining, Metallurgical and Petroleum Engineers and the Geisler Award from ASM International (Eastern New York Chapter). He is a Fellow of ASM International, the American Association for the Advancement of Science, and the Minerals, Metals and Materials Society (TMS) and a member of the National Academy of Engineering. He has a Ph.D. in metallurgy from the Rensselaer Polytechnic Institute.

BRADEN R. ALLENBY, vice president of environment, health, and safety for AT&T, was previously director of energy and environmental systems at Lawrence Livermore National Laboratory. Dr. Allenby is also vice chair of the IEEE Committee on the Environment, a member of the Advisory Committee of the United Nations Environment Programme Working Group on Product Design for Sustainability, and a former member of the Secretary of Energy's Advisory Board and the DOE Task Force on Alternative Futures for the National Laboratories. During 1992, he was the J. Herbert Holloman Fellow at the National Academy of Engineering. His expertise is in industrial ecology, especially designing for the environment and the environmental evaluation of new materials.

MALCOLM R. BEASLEY is dean of humanities and science at Stanford University. He has been professor of applied physics and electrical engineering (by courtesy) at Stanford University since 1980 and was associate professor from 1974 to 1980. Dr. Beasley was a resident fellow of engineering and applied physics at Harvard University from 1967 to 1969 and then assistant professor and associate professor from 1969 to 1974. He was awarded a B.E. in engineering physics and a Ph.D. in physics from Cornell University. He is a member of the National Academy of Sciences.

LOUIS L. BUCCIARELLI is professor of engineering and technology studies at the Massachusetts Institute of Technology, where he previously served as director of the Technology Studies Program, a predecessor to the Program in Science, Technology, and Society. His engineering research addresses problems in structural dynamics, the performance of photovoltaic solar energy systems, and energy instrumentation. He has developed software that has become the industry standard for the design of stand-alone photovoltaic systems. He has also conducted extensive ethnographic research on the interaction of engineers during the product design and development process and is the author of *Designing Engineers* (MIT Press, 1994).

JOHN V. BUSCH is president and founder of IBIS Associates, Inc. His professional focus is on economics and business development for technology-based organizations, with specialties in business development, cost modeling, and technology assessment. In addition, Dr. Busch has a technical background in materials science and engineering, industrial materials processing, and polymers and composites. He has served on the NRC Committee on Industrial Technology Assessments and the Committee to Evaluate Proposals to the New York State Science and Technology Foundation for Designation as Centers for Advanced Technology, the Panel on Intermetallic Alloy Development, and the National Materials Advisory Board.

JOHN A. DECAIRE is president of the National Center for Manufacturing Sciences, a consortium of U.S., Canadian, and Mexican corporations committed to making manufacturing in North America globally competitive through the development and implementation of next-generation manufacturing technologies. He has more than a decade of industrial experience at the Westinghouse Electric Corporation and the Raytheon Company and more than 15 years of government experience related to the development and application of advanced product and process technologies.

GEORGE E. DIETER is Glenn L. Martin Institute Professor of Engineering at the University of Maryland, having just completed a 17-year term as dean. His area of expertise is materials processing and engineering design. Professor Dieter has authored two textbooks that are widely used in the undergraduate engineering curriculum: *Mechanical Metallurgy* (McGraw-Hill, 1986 [3rd ed.])

and *Engineering Design: A Materials and Processing Approach* (McGraw-Hill, 1991 [2nd ed.]). He was awarded the A.E. White and Sauveur Award from ASM International and the Education Award from the Society of Manufacturing Engineers. He is a member of the National Academy of Engineering.

WILLIAM D. DOYLE is the MINT Chair and Professor in the Department of Physics and Astronomy at the University of Alabama and director of the Center for Materials for Information Technology. His area of research is magnetic thin films and data storage devices. He spent 31 years in industry prior to joining the University of Alabama; his last position was director of the Magnetics Division of Kodak Research Laboratories, where he was in charge of the development of heads, media, and systems. He has published more than 60 papers on magnetic materials and is a fellow of the IEEE.

NORMAN A. GJOSTEIN spent 36 years conducting and managing materials research at Ford Motor Company; he retired as director of the Materials and Manufacturing Research Laboratory in June 1996. He received a B.S. and an M.S. from the Illinois Institute of Technology and a Ph.D. in metallurgical engineering from Carnegie-Mellon University. He is an expert in the physics and chemistry of interfaces and surfaces and an authority on the application of advanced automotive materials. Dr. Gjostein is a fellow of ASM International and a member of the National Academy of Engineering.

HUGH R. MACKENZIE recently retired as group vice president of worldwide product and business-sector planning for Polaroid Corporation. He was vice president of engineering for 17 years and was responsible for product development, electronic imaging, and the design, development, and construction of manufacturing equipment and processes. He has 38 years of experience in the design of camera hardware, manufacturing equipment and processes, and assembly technologies, and he has patents and disclosures in the areas of mechanics, optics, and electronic and chemical processing. He was awarded the 1995 University of Massachusetts Outstanding Engineering Alumni Award for his achievements. He has served on advisory committees for the National Science Foundation and the American Society of Mechanical Engineers.

WILLIAM D. MANLY is a consultant at Oak Ridge National Laboratory (ORNL) for Martin Marietta Energy Systems, Inc. Mr. Manly also worked at ORNL from 1945 to 1964. He joined Union Carbide as director of materials technology in 1964, later becoming vice president and general manager of the Stellite Division. He joined Cabot Corporation when it acquired Stellite in 1970 and later became senior vice president and manager of Cabot's Engineered Materials Group. He retired from Cabot as executive vice president in 1986. He received a B.S. and an M.S.

in metallurgy from the University of Notre Dame. Mr. Manly has served on and chaired numerous committees and boards of the NRC. He is a member of the National Academy of Engineering.

NEIL E. PATON is president of Howmet Research Corporation, which is the leading supplier of premium-quality cast parts for the aerospace industry. Dr. Paton has spent approximately 35 years in the aerospace materials industry, occupying such high-level management positions as director of materials engineering and technology for Rockwell International and vice president of technology for Howmet. He is an authority on the research and development linkages of the jet-engine supply industry, as well as the technical and economic considerations that affect the application of advanced jet-engine materials.

TRESA M. POLLOCK is an associate professor in the Department of Materials Science and Engineering at Carnegie-Mellon University. Her research concerns the deformation, fracture, and processing of high-temperature structural materials, including superalloys, intermetallics, and composites. She was awarded the ASM Bradley Stoughton Award, Carnegie-Mellon George T. Ladd Research Award, National Science Foundation Presidential Young Investigator Award, and TMS High-Temperature Materials Lectureship Award.

JANE M. SHAW is research staff member and senior manager of materials and processes at the T.J. Watson Research Center of IBM. Since joining IBM, her research has been focused on new fabrication techniques, lithographic materials, polymer materials, and interconnection technology for chip and packaging applications. Her contributions to lithography include the development of photoresist modeling techniques, the fabrication of new radiation sensitive polymers, and a metallization process, silylation, which was used to fabricate all of IBMs bipolar logic chips. She has presented many invited papers, has organized and chaired sessions at international conferences, and has given short courses for the State University of New York, the University of California at Berkeley, and the American Vacuum Society. Ms. Shaw has published more than 60 papers and three book chapters and has more than 50 patents and 29 technical disclosures in the area of polymer materials and processes for the semiconductor industry. She was awarded three Outstanding Innovation Awards and a Corporate Award by IBM for materials and processes that she invented and transferred to manufacturing. In 1990, she was appointed to the IBM Academy of Technology, and in 1996 she was elected an IEEE fellow. She serves on the External Advisory Board of the National Science Foundation Science and Technology Center for High Performance Composites and Adhesives, on the Industrial Advisory Board of the Materials Processing Center at the Massachusetts Institute of Technology, and on the Industrial Advisory Board for Environmentally Conscious Materials for Michigan State University.

RONALD D. SHRIVER is vice president and plant manager of the Marysville Auto Plant for Honda of America Manufacturing, Inc. As Honda of America corporate officer, he is responsible for operations at the manufacturing plant. In addition, he is a member of the Board of Directors of Honda Engineering of America. Mr. Shriver is an internationally renowned expert in the areas of automotive manufacturing, concurrent engineering, and the implementation of new technologies in the automotive industry.

MALCOLM C. THOMAS is chief engineer of materials, processes, and life methods at Rolls-Royce Allison. Dr. Thomas has worked in aeroengine materials for 25 years in the United Kingdom and the United States. His current responsibilities include all aeroengine materials and processes. Dr. Thomas received his Ph.D. from the University of Wales and worked at International Nickel Company and GKN before joining Rolls-Royce Allison in 1986. His interests include titanium alloys, superalloys, forgings, and coatings.

ROBERT H. WAGONER is a professor in the Department of Materials Science and Engineering at Ohio State University, Columbus. Dr. Wagoner's research group investigates sheet forming from applied and fundamental perspectives, including process simulation via finite element modeling, controlled simulation tests, measurement of material formability and mechanical properties, measurement of friction, and development of plastic constitutive equations. Much of the research is conducted cooperatively with the automotive industry. He was a staff research scientist with General Motors for six years before joining the faculty at Ohio State University. He is a member of the National Academy of Engineering.